D0207581

PREDICTION AND REGULATION

APPLIED MATHEMATICS SERIES

PREDICTION AND REGULATION
by Linear Least-Square Methods
P. WHITTLE, M.Sc.(N.Z.), Ph.D.(Uppsala)

HYDRODYNAMICS AND VECTOR FIELD THEORY
Volume One: Examples in Elementary Methods
Volume Two: Examples in Special Methods
D. M. GREIG, M.A. (Cantab.), M.Sc. (Lond.), Ph.D.
T. H. WISE, M.A. (Cantab.)

INTRODUCTION TO DETERMINANTS AND MATRICES
F. BOWMAN, M.A. (Cantab.), M.Sc.Tech.

INTRODUCTION TO THE MATHEMATICS OF SERVOMECHANISMS
J. L. DOUCE, M.Sc., Ph.D.

PREDICTION AND REGULATION

by Linear Least-Square Methods

P. WHITTLE,
M.Sc.(N.Z.), Ph.D. (Uppsala)

Professor of Mathematical Statistics
University of Manchester

D. VAN NOSTRAND COMPANY, INC.
Princeton, New Jersey

Toronto New York London

D. VAN NOSTRAND COMPANY, INC.
120 Alexander St., Princeton, New Jersey (*Principal office*)
24 West 40 Street, New York 18, New York

D. VAN NOSTRAND COMPANY, LTD.
358, Kensington High Street, London, W.14, England

D. VAN NOSTRAND COMPANY (Canada), LTD.
25 Hollinger Road, Toronto 16, Canada

PRINTED IN THE UNITED STATES OF AMERICA

EDITORS' FOREWORD

The object of this series is to provide texts in applied mathematics which will cover school and university requirements and extend to the post-graduate field. The texts will be convenient for use by pure and applied mathematicians and also by scientists and engineers wishing to acquire mathematical techniques to improve their knowledge of their own subjects. Applied mathematics is expanding at a great rate, and every year mathematical techniques are being applied in new fields of physical, biological and economic sciences. The series will aim at keeping abreast of these developments. Wherever new applications of mathematics arise it is hoped to persuade a leader in the field to describe them briefly.

F. Bowman, M.A. (Cantab.), M.Sc.Tech.
Formerly Head of the Department of Mathematics,
College of Science and Technology, Manchester.

Ian N. Sneddon, M.A., D.Sc., F.R.S.E.
Simson Professor of Mathematics,
University of Glasgow.

PREFACE

The prediction of a random process in time has been studied as a topic in probability, and as a technique in the fields of communication and control: this book is an attempt at something intermediate in character.

The classical probabilistic treatment is to consider the pure prediction problem (i.e. prediction of the process which is actually observed) for a general stationary process; it culminates in a derivation of the conditions (formulae (2.7.9), (2.7.10)) for linear determinism. Several good treatments of this type already exist (Doob, 1953; Grenander and Rosenblatt, 1957); to write another would be pointless. Furthermore, from the applied point of view this approach is at once too general and too special: one will not be interested in the general stationary process, but one may well have to deal with processes which are not stationary at all, and the concept of pure prediction very soon proves over-narrow.

There is a certain amount of what I believe to be new material in the book, although in rather dispersed form. The sections on accumulated processes and multivariate processes (Chapters 8 and 9, cf Yaglom (1955, 1960)), prediction from finite samples (Chapter 7) and regulation (Chapter 10) contain some such results, which, if not new, are at least novel. The important concepts and techniques of linear least-square regulation theory are, of course, due to G. C. Newton (1952, 1957), but trying to integrate these into the book, I found that quite a number of interesting new points arose.

It was intriguing to find in the course of writing that there were two themes that recurred quite spontaneously. One of these was the characterisation of a predictor (or regulator) by the processes for which it gave an exact result. The other was the continual dualism between the time-domain and frequency-domain approaches, typified by the methods of Kolmogorov and of Wiener respectively. This dualism persisted right through to the section on regulation, where the obvious Wiener–Hopf methods had as analogue in the time-domain the fascinating idea of certainty equivalence.

Regrets which I have are that so little attention is paid to processes with irrational spectral density function and to processes in several dimensions; also that, outside the chapter on regulation, there are so few detailed discussions of special cases of the degree of complexity that would be met in practice. However, if one wishes to retain the advantages of brevity, then a fairly stern limit must be drawn.

Sections, formulae and exercises are referred to by similar conventions. Thus, "section 5" means section 5 of the current chapter, while "section (3.5)" means section 5 of chapter 3. "Ex. 3" means Exercise 3 of the current section, "Ex. (4.3)" means Exercise 3 of section 4 in the current chapter, "Ex. (6.4.3)" means Exercise 3 of section (6.4).

I shall be most grateful for any comments which readers may care to offer.

My present debts of gratitude are too numerous to list; I must recall, however, the pleasant and instructive years that I spent at the Applied Mathematics Laboratory of the New Zealand Department of Scientific and Industrial Research.

P. WHITTLE

Manchester

CONTENTS

CHAPTER 6

PROJECTION ON THE SEMI-INFINITE SAMPLE

CHAPTER 7

PROJECTION ON THE FINITE SAMPLE

CHAPTER 8

DEVIATIONS FROM STATIONARITY: TRENDS, DETERMINISTIC COMPONENTS AND ACCUMULATED PROCESSES

CHAPTER 9

MULTIVARIATE PROCESSES

CHAPTER 10

REGULATION

CHAPTER 1

INTRODUCTION

1.1 Scientific prediction

A title such as "prediction theory" is apt to raise unjustified expectations, because any claim to foretell the future, in however trivial and limited a sense, can never be regarded as something completely matter of fact. For this reason, we hasten to define our position.

Ideally, prediction is a by-product of the quantitative understanding of a situation, of a *physical model.* Thus, knowledge of Newton's laws of motion enables one to predict the paths of the planets with extreme accuracy; the laws of dynamics and elasticity enable one to predict the motion of a structure such as a bridge or a building under the influence of given applied forces. The majority of situations are *stable to specification,* in that a model which is only approximately valid will yield predictions which are also approximately valid, at least over a limited interval of time. Thus, Newton's laws do not allow for relativistic effects, and the model assumed for an engineering structure will certainly be oversimplified, but the derived predictions will not be greatly in error initially.

A model is termed *stochastic* or *deterministic* according as to whether it does or does not contain random variables. The classic models (such as the two examples above) are deterministic, and give definite predicted values. On the other hand, the future of a stochastic process is only partly determined by past values of the variables, and the idea of a definite prediction must be replaced by that of a *conditional distribution*; a probability distribution of future values, conditioned by the knowledge of past values.

Probabilistic effects may enter a model in a fundamental and inescapable fashion, as in quantum mechanics. Alternatively, they may enter because one is essentially interested in the average or "typical" behaviour of a complex system rather than in its detailed behaviour; this is the situation in statistical mechanics. Finally (and this situation is related to the last one), they may enter because there is a host of minor deterministic effects which one could not treat even if one would; the statistical effects represent this residuum of complicated and unexplained deterministic variation. The "residual" of a statistical model in econometrics, say, represents unexplained (but presumably explicable) variation of this type. (For the sake of brevity we have made these statements in a rather definite and unqualified fashion. In fact, they touch on the basic question of determinism, and are all controversial to some extent.)

Prediction based upon a hypothesis, or model, is the ideal, but it should be noted that prediction can be based upon *recognition* of a regularity as well as upon *explanation* of a regularity. Thus, one predicts the path of a tennis ball by unconsciously calling upon past experience of similar paths, rather than by calculation from the laws of motion. Similarly, one forecasts the weather more by drawing analogies from past experience than by working from a detailed physical model of the atmosphere (although this is becoming less true; experience is nowadays unified by general physical concepts, and "numerical weather forecasting" is an attempt to work almost solely from physical principles). Such forecasts are based upon the premise that, if one observes a situation similar in all relevant respects to one previously observed, then the subsequent course of events will also be similar to that previously observed. There is thus no explanation or model, but merely a compilation of possible "histories". The approach is a rather unintelligent statistical one, suffering from the defects that no unifying principles are formulated, and that the range of the relevant variables, and so the number of possible "histories", may be unmanageably large. The method is also restricted to phenomena for which any given situation will have been observed before, at least to a good approximation. That is, all situations must recur, at least in some approximate sense.

There is yet a third category of predictive statement which is sometimes made, and this is the doubtful one of *extrapolation*. For example, it was once common to make economic and demographic forecasts by fitting a straight line and extrapolating it, the assumption being that the current rate of change would continue. More generally, polynomials were fitted and extrapolated, on the assumption that some higher time-derivative would remain constant. The danger of the procedure is that it is based neither upon a physical model nor upon past experience of the same situation. For the same reason, no estimate of error can be attached to the forecast.

The fault lies not so much in the assumption that something must remain constant (for such a premise must be basic to any model) as in the arbitrary and physically unmotivated choice of what it is that is to remain constant. Nevertheless, it must be admitted that sometimes one can hardly do better: one recognises regularities in a series, but has not the information needed to construct a model, while the series does not exhibit the stationary behaviour needed to justify a direct statistical approach.

Our view-point up to now has been rather general. After this chapter we shall restrict ourselves to *linear least-square predictors*, that is, to predictors of future values which are linear functions of past values, and which are chosen so as to minimise the mean-square prediction error. These predictors will be relatively efficient if the true model is linear, and may still be useful in other cases. The technique is to some extent a blind, statistical one, although it often amounts to the virtual fitting of a linear model, which may have physical significance.

Finally, a point often overlooked is that prediction is seldom, if ever, an end in itself. One predicts the path of an enemy plane only that one may the better destroy the plane; one predicts national income in order that one may the better direct the economy. So, in the first case, the quantity to optimise will not be the accuracy of prediction but, rather, the probability of a hit—not necessarily the same thing. These considerations lead us to consider a more general problem in Chapter 10; that of optimal *regulation*.

Before specialising on linear least-square methods, we shall briefly discuss some examples of prediction from a model.

1.2 Deterministic Processes in Discrete Time

Suppose one is considering a quantity (or set of quantities) x at regular intervals of time, $t = \ldots -2, -1, 0, 1, 2, \ldots$, say. The value at time t will be denoted x_t, and the possible sequences together with their law of generation will be collectively referred to as the *process* $\{x_t\}$. If t genuinely represents physical time, then a deterministic process will in general be specified by a recursive relation of the type

$$x_t = G(x_{t-1}, x_{t-2} \ldots; t) \tag{1}$$

and a future value, $x_{t+\nu}$, can be determined from $x_{t-1}, x_{t-2}, \ldots$ by repeated use of this relation. If t is not present as a variable in G, then the process is said to be *time-homogeneous*.

Example 1: Price equilibration

Let p and d denote price and demand for a commodity, and assume, for the sake of example, that the values of these variables are determined by the relations

$$\left.\begin{aligned} p_t &= \mu_1 + p_{t-1} + \alpha d_{t-1} \\ d_t &= \mu_2 - \beta p_{t-1} - \gamma d_{t-1} \end{aligned}\right\} \tag{2}$$

This can be written in vector form

$$\mathbf{x}_t = \boldsymbol{\mu} + \mathbf{A}\mathbf{x}_{t-1} \tag{3}$$

If there is an equilibrium value, it is at

$$\bar{\mathbf{x}} = (\mathbf{I} - \mathbf{A})^{-1}\,\boldsymbol{\mu} \tag{4}$$

Suppose $\mathbf{x}_0$ is known, and one wishes to forecast $\mathbf{x}_t$. One has, from (3),

$$\begin{aligned} \mathbf{x}_t &= \sum_0^{t-1} \mathbf{A}^j\boldsymbol{\mu} + \mathbf{A}^t\mathbf{x}_0 \\ &= \bar{\mathbf{x}} + \mathbf{A}^t(\mathbf{x}_0 - \bar{\mathbf{x}}) \end{aligned} \tag{5}$$

In a case as simple as this the value of $\mathbf{A}^t$ would be calculated from the spectral representation of $\mathbf{A}$. The qualitative behaviour of $\mathbf{x}_t$ will depend upon the nature of the eigenvalues, the zeros of $|\lambda\mathbf{I} - \mathbf{A}|$:

$$\lambda = \frac{1 - \gamma \pm \sqrt{(1+\gamma)^2 - 4\alpha\beta}}{2} \tag{6}$$

These must be less than unity in modulus if the system is to be stable, and ultimately reach equilibrium. If they are complex, then the system will show oscillatory tendencies.

Example 2: Population projection (see Leslie, (1945), (1948)).

In order to make the problem discrete, we take a year as the unit of time, and group the population by age in years, so that n_{jt} is the number of people in year t who are in their jth year of life. Let θ_{jt} and ϕ_{jt} be the yearly fertility and survival rates for individuals in this class, so that a plausible deterministic model would be

$$\left.\begin{aligned} n_{1t} &= \sum_{j=1}^{\infty} \theta_{j,t-1} n_{j,t-1} \\ n_{jt} &= \phi_{j-1,t-1} n_{j-1,t-1}. \quad (j = 2, 3 \ldots) \end{aligned}\right\} \tag{7}$$

(It is conventional to ignore distinctions between sexes; to introduce rates of mating would make the model non-linear.) The equation system (7) may be written in vector form

$$\mathbf{n}_t = \mathbf{A}_t \mathbf{n}_{t-1} \tag{8}$$

If the age structure of the population is known at time $t = 0$, then the predicted future state is obtained by calculating

$$\mathbf{n}_t = \mathbf{A}_t \mathbf{A}_{t-1} \mathbf{A}_{t-2} \ldots \mathbf{A}_1 \mathbf{n}_0 \tag{9}$$

This is the kind of model actually used for "population projection", future values of the rates θ_{jt} and ϕ_{jt} being very commonly obtained by extrapolation with respect to t of trend-curves fitted to observed values. If the system were time-homogeneous, so that these rates did not depend upon t then one would have

$$\mathbf{n}_t = \mathbf{A}^t \mathbf{n}_0 \tag{10}$$

Ex. Show that in the time-homogeneous case the eigenvalues λ of $\mathbf{A}$ are determined by

$$\sum_{j=1}^{\infty} \lambda^{-j} \theta_j \prod_{k=1}^{j-1} \phi_k = 1$$

and that if λ_1 is the greatest eigenvalue, then ultimately the population multiplies itself by a factor of λ_1 every year, and reaches an age distribution in which the number in the jth year of life is proportional to $\lambda_1^{-j} \prod_{k=1}^{j-1} \phi_k$.

A case of particular importance is that in which (1) reduces to a simple recursion

$$x_t = G(x_{t-1}) \tag{11}$$

so that x_t constitutes a "complete description" of the state of the process at time t. Our two examples were both of this form; or rather, the variables were so chosen that they could be cast in this form.

1.3 Deterministic Processes in Continuous Time

Let us again consider a process $\{x_t\}$, where x_t may be a vector (or even denote a continuously infinite set of variables; see the second example below) but t can now vary continuously. Then the analogue of relation (2.1) would be

$$x_t = G(x_s, s < t; t) \tag{1}$$

where G is some kind of functional of past values. A *formally* equivalent specification would be

$$\frac{dx_t}{dt} = F(x_s, s < t; t) \tag{2}$$

where F is a related functional. In physical examples the form (2) is likely to be preferable, as the functional F will be better behaved.

As before, one can specialise to the time-homogeneous case; also to the case in which x_t constitutes a "complete description", so that x_t is determined by x_{t-} alone. An important case is

$$\frac{dx_t}{dt} = \mathbf{A}x_t \tag{3}$$

where $\mathbf{A}$ is a linear operator. This has the formal solution (analogous to (2.10))

$$x_t = e^{t\mathbf{A}}x_0 = \left[\sum_0^\infty \frac{(t\mathbf{A})^j}{j!}\right]x_0 \tag{4}$$

Example 1: Kinetics of a chemical reaction

Suppose that compounds 1, 2 and 3 are present in amounts m_1, m_2 and m_3 (or $m_1(t)$, etc., if one wishes to be more specific), and that there is a reaction

$$1 + 2 \rightleftharpoons 3 \tag{5}$$

whose rate is governed by the equations

$$\left.\begin{aligned} \frac{dm_3}{dt} &= \lambda m_1 m_2 - \mu m_3 \\ &= -\frac{dm_1}{dt} = -\frac{dm_2}{dt} \end{aligned}\right\} \tag{6}$$

If the initial amounts ($t = 0$) of substances 1 and 2 were M_1 and M_2, then (6) can be rewritten

$$\begin{aligned} \frac{dm_3}{dt} &= \lambda(M_1 - m_3)(M_2 - m_3) - \mu m_3 \\ &= \lambda(\alpha - m_3)(\beta - m_3) \end{aligned} \tag{7}$$

say. If $m_3(0) = 0$ then we find from (7) that

$$m_3(t) = \frac{\alpha\beta({}^{\lambda\alpha t} - e^{\lambda\beta t})}{\alpha e^{\lambda\alpha t} - \beta e^{\lambda\beta t}} \longrightarrow \min(\alpha, \beta) \tag{8}$$

and this gives the predicted course of the reaction. This is the first non-linear model we have considered: few are so tractable.

Example 2: Temperature fluctuations in the earth

Suppose that the temperature x is dependent only upon time, t, and depth, ξ; let it be denoted $x(\xi, t)$. On conventional assumptions one has

$$\left.\begin{aligned} \frac{\partial x}{\partial t} &= \frac{\kappa^2}{2}\frac{\partial^2 x}{\partial \xi^2} \quad (\xi > 0) \\ x &\longrightarrow 0 \quad (\xi \longrightarrow \infty) \end{aligned}\right\} \tag{9}$$

From (9) one can derive a solution in terms of the surface temperature:

$$x(\xi, t) = \frac{\xi}{\sqrt{2\pi\kappa}} \int_0^\infty e^{-\xi^2/(2\kappa s)} x(0, t-s) s^{-\frac{3}{2}} ds \tag{10}$$

This is a "prediction", not in time, but in the spatial co-ordinate ξ. If the Fourier transforms

$$\bar{x}(\xi, \omega) = \int_{-\infty}^{\infty} e^{-i\omega t} x(\xi, t) dt \tag{11}$$

exist, then (10) can be written

$$\bar{x}(\xi, \omega) = e^{-(1-i)\xi|\omega/u|^{\frac{1}{2}}} \bar{x}(\xi, 0) \tag{12}$$

1.4 Stochastic Processes in Discrete Time

The analogue of the recursion (2.1) will now be the specification of a conditional probability

$$P(x_t \in \mathscr{A} | x_{t-1}, x_{t-2}, \ldots) = Q(\mathscr{A}; x_{t-1}, x_{t-2}, \ldots; t) \tag{1}$$

where $P(x_t \in \mathscr{A} | y)$ denotes the probability that x_t takes a value in a set $\mathscr{A}$, conditional on given values for the random variables y. If the function Q is known for a suitably large class of sets $\mathscr{A}$, then from (1) one can calculate the probability of a variety of events defined on the future, and in particular can calculate $P(x_{t+\nu} \in \mathscr{A} | x_{t-1}, x_{t-2} \ldots)$, at least in principle.

If t does not appear explicitly in Q, then the process is *time-homogeneous*; if x_{t-1} is the only x appearing then the process is *Markov* (and x again supplies a "complete description" of the state of the process, in an extended sense).

As an example, suppose that the model (2.2) is modified to

$$\mathbf{x}_t = \boldsymbol{\mu} + \mathbf{A}\mathbf{x}_{t-1} + \boldsymbol{\epsilon}_t \tag{2}$$

where the vectors $\boldsymbol{\epsilon}_t$ are independently and normally distributed, and

$$\left.\begin{aligned} E(\boldsymbol{\epsilon}_t) &= 0 \\ E(\boldsymbol{\epsilon}_t\boldsymbol{\epsilon}'_t) &= V \end{aligned}\right\} \tag{3}$$

From these facts one could calculate the probability (1). However, in this case it is easier to form some idea of the behaviour of the process by "solving" (2), so that, analogously to (2.5),

$$\mathbf{x}_t - \bar{\mathbf{x}} = \sum_{s=0}^{t-1} \mathbf{A}^s \boldsymbol{\epsilon}_{t-s} + \mathbf{A}^t(\mathbf{x}_0 - \bar{\mathbf{x}}) \tag{4}$$

For a given $\mathbf{x}_0$ we find from (4) that the expected value of $\mathbf{x}_t$ is given by expression (2.5), so that this is still a "prediction" in a certain sense. The vector "prediction error" is $\boldsymbol{\delta}_t = \sum_0^{t-1} \mathbf{A}^s \boldsymbol{\epsilon}_{t-s}$, with covariance matrix

$$E(\boldsymbol{\delta}_t\boldsymbol{\delta}'_t) = \sum_0^{t-1} \mathbf{A}^s \mathbf{V}(\mathbf{A}')^{s-1} \tag{5}$$

As a second example, consider a stochastic version of the population model specified by equations (2.7). The natural way to "stochasticise" the model is to re-interpret θ_{jt} and ϕ_{jt} as *probabilities* that any single individual of the n_{jt} in the appropriate class reproduces or survives during the year, distinct individuals having statistically independent histories. If we suppose this, and introduce the probability generating function,

$$\Pi_t(z_1, z_2, \ldots) = E[\prod_{j=1}^{\infty} z_j^{n_{jt}}] \tag{6}$$

then instead of (2.6) we have a relation

$$\Pi_t(z_1, z_2, \ldots) = \Pi_{t-1}(\zeta_1, \zeta_2, \ldots) \tag{7}$$

where

$$\zeta_j = [1 + \phi_{j,t-1}(z_{j+1} - 1)][1 + \theta_{j,t-1}(z_1 - 1)] \tag{8}$$

For a small, isolated, unisexual population this would be quite a plausible model. For a human population the stochastic effects built into the model would certainly be present, but would be secondary in comparison with effects due to immigration, social and economic changes, etc.

Equations (7) and (8) effectively specify the probabilities (1). In principle, one could thus determine the distribution of $\mathbf{n}_t$ conditional on $\mathbf{n}_0$, say. In fact, this is not so simple, although one can relatively easily determine the means and covariances:

$$\begin{aligned} \boldsymbol{\mu}_t &= E(\mathbf{n}_t) \\ \mathbf{V}_t &= E(\mathbf{n}_t - \boldsymbol{\mu}_t)(\mathbf{n}_t - \boldsymbol{\mu}_t)' \end{aligned} \tag{9}$$

One finds from (7) that these obey the recurrences

$$\boldsymbol{\mu}_t = \mathbf{A}_t \boldsymbol{\mu}_{t-1} \tag{10}$$

$$\mathbf{V}_t = \mathbf{A}_t \mathbf{V}_{t-1} \mathbf{A}'_t + \begin{bmatrix} \sum_j \theta_{j,t-1}(1-\theta_{j,t-1})\mu_{j,\,t-1} & \cdot & \cdot & \cdot \\ \cdot & \phi_{1,t-1}(1-\phi_{1,t-1})\mu_{1,t-1} & \cdot & \cdot \\ \cdot & \cdot & \phi_{2,t-1}(1-\phi_{2,t-1})\mu_{2,t-1} & \cdot \\ \cdot & \cdot & \cdot & \cdot \end{bmatrix}$$

where $\mathbf{A}_t$ is the matrix defined by equations (2.6) and (2.7). Thus, the expected future course of the process follows the prediction from the deterministic model. The variances of the prediction errors can be calculated from (11).

1.5 Stochastic Processes in Continuous Time

The analogue of the equation (or equation system) (2.1) will now be

$$\frac{d}{dt} P(x_t \in \mathscr{A} | x_s, s < t) = R(\mathscr{A}; x_s, s < t; t) \tag{1}$$

If x_s appears in the functional R only for $s = t-$, then the process is *Markov*. If t does not appear explicitly in R, then the process is *time-homogeneous*. In many cases a relation of type (1) may hold only in a formal sense. Also, another formulation may be more convenient: in terms of the variables themselves (cf. equation (4.2)) or of generating functions (cf. equation (4.7)). However, in principle relations of type (1) exist, and can be used to determine probabilities of events defined on the future.

As an example, consider a stochastic version of the chemical reaction model specified by equations (3.6). One might modify these equations by the addition of a "residual" ε_t, as (2.3) was modified to (4.2). This would be plausible if extraneous variation were being introduced into the situation (e.g. fluctuations of temperature or of mixing). Another method, reasonable if the amounts of the reactants are so small that fluctuations due to the discreteness of molecular structure must be taken into account, would be to re-interpret $\lambda m_1 m_2$ and μm_3 as *probability intensities of transition* rather than as rates.

With these assumptions, the stochastic analogue of equations (3.6) becomes

$$\frac{\partial \Pi}{\partial t} = (z_3 - z_1 z_2)\left[\lambda \frac{\partial^2 \Pi}{\partial z_1 \partial z_2} - \mu \frac{\partial \Pi}{\partial z_3}\right] \tag{2}$$

where π is the probability generating function

$$\Pi(z_1, z_2, z_3; t) = E[\prod_1^3 z^{m_i(t)}] \tag{3}$$

and $m_i(t)$ is the actual number of molecules of substance i at time t. If one takes account of the restrictions

$$m_i = M_i - m_3, \qquad (i = 1, 2) \tag{4}$$

then (2) can be rewritten as an equation for a probability generating function with a single argument, just as (3.6) could be rewritten as (3.7).

In fact, neither of these two stochastic versions of the reaction process can be solved to yield conditional distributions of future values in closed form. Non-linear processes seem to be even less tractable in the stochastic than in the deterministic case, and constitute a standing problem.

As a second example, consider the heat-diffusion problem of equations (3.9), but suppose that the "input" function, $x(0, t)$, assumed known, is random. Then the solutions (3.10) and (3.12) still hold, at least if the integrals and integral transforms defined converge in some suitable sense. From the explicit solution (3.10) one can determine the statistical properties of the random function $x(\xi, t)$. For instance, if $\{x(\xi, t)\}$, as a process *in time*, is stationary with spectral density function $f_\xi(\omega)$, then it follows from (3.12) that

$$f_\xi(\omega) = e^{-2\xi|\omega/\kappa|^{\frac{1}{2}}} f_0(\omega) \tag{5}$$

1.6 Linear Least-square Prediction and Estimation

While the ideal stochastic prediction would be the complete specification of a conditional distribution, this is seldom practicable. From this point onwards we shall restrict ourselves to consideration of the *linear least-squares (l.l.s.) predictor*, for which $x_{t+\nu}$ is estimated by a linear function of known values,

$$\hat{x}_{t+\nu} = \mu + \sum_j \gamma_j x_{t-j} \tag{1}$$

the coefficients μ, γ_j being chosen on the criterion that the *prediction mean square error (m.s.e.)*

$$E(\delta^2) = E(\hat{x}_{t+\nu} - x_{t+\nu})^2 \tag{2}$$

be a minimum. The sum in (1) ranges over all times at which x has been observed. In the conventional prediction problem one will have observed the values $x_t, x_{t-1}, \ldots$, to that j will range from 0 to $+\infty$.

Suppose that the $\{x_t\}$ process is stationary, and contains no deterministic components (see section (2.6)). It will thus have zero mean value, since a non-zero mean would be regarded as a deterministic component. One easily sees that the constant term μ in (1) should be zero under these circumstances, so that in the conventional prediction problem we should have simply

$$\hat{x}_{t+\nu} = \sum_0^\infty \gamma_j x_{t-j} \tag{3}$$

In this case, all that one need know of the $\{x_t\}$ process in order to determine the γ_j from the least-square principle above is its autocovariance function:

$$\Gamma_s = \operatorname{cov}(x_t, x_{t-s}) \tag{4}$$

This is just about the minimum information on the process that will allow one to construct a non-trivial predictor.

The l.l.s. approach is theoretically elegant, in that it fits in naturally with the representation theory of stationary processes; concerned with canonical linear representations calculated from the covariances Γ_s. It is also a rather natural approach physically, since, in continuous time at least, the calculation of the predictors will be performed by electrical or mechanical filters, many of whose basic components will behave linearly.

A generalisation of the stationary, purely non-deterministic process would be that in which x_t could be represented

$$x_t = \chi_t + \eta_t \tag{5}$$

where $\{\chi_t\}$ is deterministic, and $\{\eta_t\}$ purely non-deterministic and stationary. If χ_t is known, then it can be subtracted from x_t, and the residual η_t predicted as before. Thus, one effectively has a predictor of the form (1), but with μ a known function of t. In fact, χ_t will seldom be known, but a common assumption is to represent it

$$\chi_t = \sum_j \beta_j g_j(t) \tag{6}$$

where the $g_j(t)$ are known functions, but the coefficients β_j are unknown. The β_j can themselves be estimated by l.l.s. methods (see sections (4.3) and (8.3)), and the actual prediction formula again falls into the pattern (3). To deal with more general types of non-stationarity by these methods is seldom practicable (see section (8.2)), although one can deal with processes obtained by simple linear operations on stationary processes (see section (8.5)).

It is interesting to note that the *unrestricted* (and so possibly non-linear) least square predictor is the conditional mean

$$\check{x}_{t+\nu} = E(x_{t+\nu} | x_t, x_{t-1}, \ldots) \tag{7}$$

If the $\{x_t\}$ process is normal, then expression (7) does in fact reduce to the l.l.s. predictor (see Ex. (4.1.6)).

Of course, the l.l.s. predictor will not in general be as precise as a less restricted predictor, which uses more information on the process. Indeed, we shall later (section (2.6)) give examples of processes which are deterministic in the sense of this chapter, and yet for which the m.s.e. of the linear predictor is non-zero. This is to be expected: l.l.s. prediction is nevertheless a useful technique, and certainly demands the minimum amount of information on the process.

The nearer the process itself is to being linear (i.e. generated by linear relations), the nearer the l.l.s. estimate is to being fully efficient. Indeed, if the model is truly linear, then the l.l.s. procedure will often (although not always) recover the actual relations of the model, so that the technique becomes more than a blind one of l.s. approximation. For example, in the case of model (4.2), if the ϵ_t are uncorrelated, then the l.l.s. prediction will be just the relation (2.5), with prediction error given by (4.5). This will not be so if the ϵ_t are auto-correlated, because then there is extra information to be gained from the auto-correlation.

The introduction of the l.l.s. idea leads naturally to other operations than that of prediction. For example, one could perhaps *choose* the coefficient α in relation (2.2) in such a way as to minimise the mean square variation of price in model (4.2). This is an elementary example of l.s. *regulation*; we shall study more sophisticated examples and techniques in the final chapter.

Again, if one can use a sample $(x_t, x_{t-1}, \ldots)$ to predict $x_{t+\nu}$ by l.s. methods one can also use it to estimate other quantities—the value of a term in some related time series, perhaps. In order to cover this extended application we shall often speak of *estimation* rather than prediction, and refer to the *linear least-square estimator* (*l.l.s.e.*).

The sample of values extending from the infinite past to the present, denoted $(x_t, x_{t-1}, \ldots)$ or $(x_s; s \leqslant t)$, will be referred to as a *semi-infinite sample* (with an obvious analogue in continuous time). Sometimes one has only a *finite sample* $(x_0, x_1, \ldots x_{n-1})$, and there are occasions where one can plausibly assume knowledge of the *infinite sample*: (x_t, all integral t) in the discrete time case, and (x_t, all real t) in the continuous time case. The method of finding the l.l.s.e. varies from case to case; we shall attempt to make the distinctions clear in Chapters 5–7.

The theory of l.l.s. prediction for stationary time series was developed simultaneously by Wiener (1949) and Kolmogorov (1939, 1941b). Their methods differed: this will be made clear later in the book, as we use both methods. Wiener's work was more directed towards applications; the particular one important at the time being the prediction of aircraft flight paths, for purposes of fire control. The field of application has broadened greatly since then, however, as will be clear from Chapter 10, where we shall concentrate most discussion of actual physical situations. In particular, the theory, modified and extended in various ways, constitutes a tool of considerable potential value for forecasting and regulation in industry, and in the economy generally.

We shall be working with stationary processes constantly throughout the book, and so we summarise the relevant theory in the next chapter.

CHAPTER 2

STATIONARY AND RELATED PROCESSES

Conventional prediction theory is closely bound up with the ideas of stationary and of linear processes. This chapter is a summary (*not* an exposition!) of such of these ideas as are necessary to the understanding of the rest of the book. At the same time we fix our notation.

2.1 Notation

Abbreviations which we shall use in the text are: l.s., "least square"; l.l.s., "linear least square"; l.l.s.e., "linear least square estimate"; m.s., "mean square"; m.s.e., "mean square error"; a.r., "autoregression"; m.a., "moving average"; s.d.f., "spectral density function"; and s.d.m., "spectral density matrix".

Very often, although not invariably, the same letter will be used to denote a function and the coefficients in its Laurent expansion:

$$\gamma(z) = \sum_{-\infty}^{\infty} \gamma_j z^j \tag{1}$$

If the region in which this expansion is valid is understood (and with only one exception, equation (10.5.27), the regions we discuss always include $|z| = 1$) we shall use the notation

$$[\gamma(z)]_m^{(n)} = \sum_m^n \gamma_j z^j \tag{2}$$

If the lower subscript m is omitted it is understood to be $-\infty$; similarly, if n is omitted it is understood to be $+\infty$. For $m = 0$ we shall use a particular notation:

$$[\gamma(z)]_+ = [\gamma(z)]_0 = \sum_0^{\infty} \gamma_j z^j \tag{3}$$

$$[\gamma(z)]_- = [\gamma(z)]^{(-1)} = \sum_{-\infty}^{-1} \gamma_j z^j \tag{4}$$

Our arguments so often concern real γ_j, and z on $|z| = 1$, that we are scarcely guilty of inconsistency if we use $\overline{\gamma(z)}$ (or sometimes simply $\bar{\gamma}$) to denote $\gamma(z^{-1})$, and $|\gamma(z)|^2$ (or sometimes $|\gamma|^2$) to denote $\gamma(z)\gamma(z^{-1})$. Correspondingly, if $\boldsymbol{\gamma}(z)$ is a matrix function of the scalar z, the symbol $\boldsymbol{\gamma}(z)^\dagger$ will denote the transpose of $\boldsymbol{\gamma}(z^{-1})$. This notation will be extended to operators in section (10.1).

We shall use $\mathscr{A}$ to denote "absolute term in Laurent expansion on $|z| = 1$", so that, for example,

$$\mathscr{A}\gamma(z) = \frac{1}{2\pi i}\oint_{|z|=1} \gamma(z)\frac{dz}{z} = \gamma_0 \tag{5}$$

$$\mathscr{A}|\gamma(z)|^2 = \Sigma\gamma_j{}^2 \tag{6}$$

if (1) is valid on $|z| = 1$.

The conventions for the continuous time case (where one deals with Fourier integrals rather than with Fourier or Laurent series) are analogous, if one makes the correspondence,

$$z = e^{-i\omega} \tag{7}$$

Thus, if a function has a valid Fourier expansion

$$\gamma(\omega) = \int_{-\infty}^{\infty} \gamma_s e^{-i\omega s} ds \tag{8}$$

(generally to be understood for real ω, at least), then

$$[\gamma(\omega)]_\lambda^{(\mu)} = \int_{\lambda-}^{\mu+} \gamma_s e^{-i\omega s} ds \tag{9}$$

with the same convention on omitted λ or μ as before. Corresponding to (3) and (4)

$$[\gamma(\omega)]_+ = [\gamma(\omega)]_0 = \int_{0-}^{\infty} \gamma_s e^{-i\omega s} ds \tag{10}$$

$$[\gamma(\omega)]_- = [\gamma(\omega)]^{0-} = \int_{-\infty}^{0-} \gamma_s e^{-i\omega s} ds \tag{11}$$

By $\bar{\gamma}$, $|\gamma|^2$ and $\boldsymbol{\gamma}^\dagger$ we shall mean $\gamma(-\omega)$, $\gamma(\omega)\gamma(-\omega)$ and $\gamma(-\omega)'$. (It will always be clear from the context whether the discussion concerns discrete or continuous time.) Corresponding to (5),

$$\mathscr{A}\gamma(\omega) = \frac{1}{2\pi}\int_{-\infty}^{\infty} \gamma(\omega) d\omega = \gamma_0 \tag{12}$$

A stationary process in time with variate x_t will be denoted $\{x_t\}$; we use the same notation for both discrete and continuous cases. If the variate is a vector $\mathbf{x}_t$, then we denote the *autocovariance matrix of lag s* by

$$\mathbf{\Gamma}_s = E(\mathbf{x}_t \mathbf{x}^\dagger{}_{t-s}) \tag{13}$$

We consider the case of a complex vector variate in order to avoid repetition: the real and scalar cases are obviously included. The only differentiation in notation is that we shall not use bold-face characters for scalar quantities. The symbol E denotes the expectation operator.

In the discrete time case we shall have cause to consider the auto-covariance generating function (or matrix), defined, if the series converges, by

$$\mathbf{g}(z) = \sum \mathbf{\Gamma}_s z^s \tag{14}$$

This is an immediate but convenient modification of the s.d.f. (scalar case) or s.d.m. (vector case)

$$\mathbf{f}(\omega) = \mathbf{g}(e^{-i\omega}) = \sum \mathbf{\Gamma}_s e^{-i\omega s} \tag{15}$$

The corresponding quantity in continuous time (also termed the s.d.f. or s.d.m.) is

$$\mathbf{f}(\omega) = \int_{-\infty}^{\infty} \mathbf{\Gamma}_s e^{-i\omega s} ds \tag{16}$$

For the cases where expressions (14)–(16) do not converge one must employ the *spectral distribution function* (or *spectral distribution matrix*, in the vector case)

$$\mathbf{F}(\omega) = \text{const.} + \lim_{n\to\infty} \sum_{-n}^{n} \left(\frac{e^{-i\omega s} - 1}{-i\omega} \right) \mathbf{\Gamma}_s \tag{17}$$

(discrete time)

$$\mathbf{F}(\omega) = \text{const.} + \lim_{t\to\infty} \int_{-t}^{t} \left(\frac{e^{-i\omega s} - 1}{-i\omega} \right) \mathbf{\Gamma}_s ds \qquad (18$$

(continuous time)

We shall sometimes consider two jointly stationary vector processes, $\{\mathbf{x}_t\}$ and $\{\mathbf{y}_t\}$, and shall then write, analogously to (15),

$$\mathbf{f}_{xy}(\omega) = \mathbf{g}_{xy}(e^{-i\omega}) = \sum_s \mathbf{\Gamma}_s^{(xy)} e^{-i\omega s} = \sum_s E(\mathbf{x}_t \mathbf{y}^\dagger_{t-s}) e^{-i\omega s} \tag{19}$$

for the cases when this series converges. The analogous definitions for continuous time and for $\mathbf{F}_{xy}(\omega)$ will be clear. The quantity defined in (15) could thus be written $\mathbf{f}_{xx}(\omega)$, or $\mathbf{g}_{xx}(z)$, but it will often not be necessary to be this explicit.

We shall occasionally have use for the translation and differentiation operators, U and D, with the properties

$$Ux_t = x_{t-1} \tag{20}$$

$$Dx_t = \frac{d}{dt} x_t \tag{21}$$

2.2 Linear Operations: Response and Transfer Functions

We shall consider first the discrete time case. Consider the system of equations

$$L(x_t) = u_t \tag{1}$$

where t ranges over all integral values. Here L is a constant coefficient linear operator, and so can be represented in the form

$$L(x_t) = A(U)x_t = \sum_j a_j x_{t-j} \tag{2}$$

The equation system (1) can be regarded as describing the action of a linear mechanism, or filter, u_t being the *input*, x_t the *output*, and L being an operator characteristic of the mechanism. The problem is to invert (1), and obtain x_t in terms of u_t.

Consider the input consisting of a unit value at $t = 0$:

$$u_t = \delta_t = \begin{cases} 1, & t = 0 \\ 0, & t \neq 0 \end{cases} \tag{3}$$

If t really represents physical time, then we can assume in (2) that $a_j = 0$ $(j < 0)$, $a_0 \neq 0$, and so regard (1) as a recursion determining x_t in terms of $x_{t-1}, x_{t-2}, \ldots$ and $u_t, u_{t-1}, \ldots$ The solution of (1), (3), under the conditions $x_t = 0$ $(t < 0)$ is thus determinate, and we have

$$x_t = b_t, \qquad (t \geqslant 0) \tag{4}$$

say. The sequence b_j is known as the *transient response function* of the mechanism; it represents the response to a pulse input. If

$$B(z) = \sum_0^\infty b_j z^j \tag{5}$$

then

$$A(z)B(z) = 1 \tag{6}$$

and this relation can be used to determine the b_j sequence. Relation (6) is valid even if $A(z)$ converges for no z, in the sense that the system of equations implied by (6),

$$\sum_{k=0}^{j} a_k b_{j-k} = \delta_j \qquad (j \geqslant 0) \tag{7}$$

always determines the b's correctly.

It follows from (3), (4) and the linearity of L that a solution of (1) for a general input u_t is

$$x_t = \sum_0^\infty b_j u_{t-j} \tag{8}$$

To this solution can be added any solution of the homogeneous equation

$$L(x_t) = 0 \tag{9}$$

This indeterminacy will be removed if one can specify initial values for the sequence x_t.

The initial value problem can be solved explicitly by these methods. Suppose that x_t is specified for $t < \tau$, and let us define the generating functions

$$\left.\begin{aligned} u(z) &= \sum_{\tau}^{\infty} u_t z^t \\ x(z) &= \sum_{\tau}^{\infty} x_t z^t \\ x_-(z) &= \sum_{-\infty}^{\tau-1} x_t z^t \end{aligned}\right\} \tag{10}$$

Then, if (1) holds for $t \geqslant \tau$ we find that

$$A(z)[x(z) + x_{-}(z)] = u(z) + \ldots \tag{11}$$

where the dots indicate terms in z^t $(t < \tau)$. Discarding these terms, we have then

$$A(z)x(z) + [A(z)x_{-}(z)]_{\tau} = u(z) \tag{12}$$

or

$$x(z) = \frac{u(z)}{A(z)} - \frac{1}{A(z)}[A(z)x_{-}(z)]_{\tau} \tag{13}$$

which determines the unknown x values. Again, this relation is true irrespective of questions of convergence, in the sense that formal expansion of $A(z)^{-1}$ in non-negative powers of z, and equation of coefficients of powers of z, will yield the correct x_t solutions.

If $x_t = y^{-t}$ $(t < \tau)$ then (13) becomes

$$x(z) = \frac{u(z)}{A(z)} - \frac{y^{-\tau+1}z^{\tau}}{A(z)}\left[\frac{A(z) - A(y)}{z - y}\right] \tag{14}$$

This expression is a double generating function: the coefficient of $z^t y^{-s}$ $(t \geqslant \tau, s < \tau)$ being the coefficient of x_s in the solution for x_t.

The first term in (13), $u(z)/A(z)$, corresponds to the solution (8). If we can show that the contribution from the final term in (13) becomes negligible as τ recedes to $-\infty$, then (8) will be *the* solution to the equation system (1). A sufficient condition for the validity of this conclusion is that the sequence u_t should have a Fourier representation

$$u_t = \int_{-\pi}^{\pi} e^{i\omega t} dK(\omega) \tag{15}$$

where $K(\omega)$ is a function of bounded variation in $(-\pi, \pi)$ and that

$$\sum_{0}^{\infty} |b_j| < \infty \tag{16}$$

Condition (16) is in the nature of a *stability condition* on the mechanism corresponding to L. If $A(z)$ is a polynomial, then the condition is equivalent to the demand that all the zeros of $A(z)$ lie outside the unit circle. In general, condition (16) will ensure that $A(z)^{-1}$ is regular in $|z| \leqslant 1$.

If (15) and (16) hold, then solution (8) can be written in the same form as (15):

$$x_t = \int_{-\pi}^{\pi} e^{i\omega t} B(e^{-i\omega})\, dK(\omega) = \int_{-\pi}^{\pi} e^{i\omega t} \frac{dK(\omega)}{A(e^{-i\omega})} \tag{17}$$

This gives another, and extremely powerful, insight on the system (1). Relations (8) and (17) are solutions of (1) derived respectively in the *time domain* and in the *frequency domain.* The sequence b_j is the response to a simple pulse, (3); the sequence

$$x_t = \frac{e^{i\omega t}}{A(e^{-i\omega})} \tag{18}$$

is the response to a simple sinusoid; $u_t = e^{i\omega t}$. The function $B(e^{-i\omega}) = A(e^{-i\omega})^{-1}$ is termed the *frequency response function* or the *transfer function* of the system; we shall in general use the latter term.

All these results have immediate analogues in continuous time. If (1) now holds for all real t, then, corresponding to (8), a formal solution is

$$x_t = \int_0^\infty b_s u_{t-s}\, ds \tag{19}$$

where b_s is again the transient response function, the response to a delta-function input. If

$$L(e^{i\omega t}) = A(\omega)e^{i\omega t} \tag{20}$$

$$B(\omega) = \int_0^\infty b_s e^{-i\omega s}\, ds \tag{21}$$

then, corresponding to (6),

$$A(\omega)B(\omega) = 1 \tag{22}$$

A sufficient condition for (19) to be the appropriate solution of (1) ((1) being understood to hold for all t) is that

$$\int_{-\infty}^\infty |b_s|\, ds < \infty \tag{23}$$

and that u_t be representable

$$u_t = \int_{-\infty}^\infty e^{i\omega t}\, dK(\omega) \tag{24}$$

where $K(\omega)$ is a function of bounded variation. Solution (19) can then alternatively be written

$$x_t = \int_{-\infty}^\infty e^{i\omega t} B(\omega)\, dK(\omega) = \int_{-\infty}^\infty e^{i\omega t} \frac{dK(\omega)}{A(\omega)} \tag{25}$$

Note that in (20) we have *defined* $A(\omega)$ as the inverse of the transfer function, rather than as the Fourier transform of a weight function corresponding to the coefficients a_j. The point is that if L involves differential operators, say, then no proper weight function exists, unless one is prepared to admit the Dirac delta function and its derivatives as proper functions in some sense. The best course is to regard $L(x_t)$ as a linear functional of the function x_t, and to regard $A(\omega)$ as being defined by this functional for the special case $x_t = e^{i\omega t}$.

An important special case is that in which L is a constant-coefficient differential operator, so that

$$L = P(D) \tag{26}$$

where P is a polynomial. One finds then that

$$A(\omega) = P(i\omega) \tag{27}$$

and the condition (23) is equivalent to the demand that all the zeros of P have negative real part (or that all the zeros of $A(\omega)$ have positive imaginary part).

An analogue of the initial value solution (13) holds, without restriction on $A(\omega)$, although some care is needed in application of the $[\ \ ]_+$ operator. We shall record the result only for the special case (26):

$$\int_\tau^\infty e^{-\lambda t} x_t dt = \frac{1}{P(\lambda)} \int_\tau^\infty e^{-\lambda t} u_t dt - \frac{e^{-\lambda \tau}}{P(\lambda)} \left[\frac{P(D) - P(\lambda)}{D - \lambda} x_t \right]_{t=\tau} \tag{28}$$

It should be noted that all these results extend immediately to the case of vector variates, and matrices of operators.

2.3 The Spectral Representation of a Stationary Process (Scalar Case)

We shall be interested in linear systems of the type (2.1) for the case when the input sequence u_t constitutes a stationary process, either in discrete or in continuous time. The results of section (2) will still hold formally in this case, but one must re-examine all questions of convergence, since one must now think in terms of some type of statistical convergence. In general, one adopts the criterion of mean-square convergence, as being a natural and convenient one for linear representations.

An important result is, that if the process $\{u_t\}$ is stationary in the wide sense (i.e. $E(u_t)$ and $\text{cov}\,(u_t, u_{t-s})$ exist and are functionally independent of t), then a representation exists of the types (2.15), (2.24):

$$u_t = \frac{1}{\sqrt{2\pi}} \int e^{i\omega t}\, d\zeta(\omega) \tag{1}$$

(Cramer, 1942).

The limits of integration are $\pm\pi$ if t takes integral values, $\pm\infty$ if t takes all real values. Relation (1) is the *spectral representation of the variate* u_t; properly, it represents u_t only in mean square, i.e. the difference of the two members of equation (1) has an expected square modulus of value zero.

The function $\zeta(\omega)$ must obviously also be random. If this were all that were true, the representation (1) might not be so remarkable; one can add, however, that the process $\zeta(\omega)$ is *orthogonal*, i.e. it has uncorrelated increments:

$$E[\zeta(\omega_2) - \zeta(\omega_1)]\overline{[\zeta(\omega_4) - \zeta(\omega_3)]} = 0 \tag{2}$$
$$(\omega_1 < \omega_2 < \omega_3 < \omega_4)$$

Furthermore

$$E|\zeta(\omega_2) - \zeta(\omega_1)|^2 = F(\omega_2) - F(\omega_1) \qquad (\omega_1 < \omega_2) \tag{3}$$

where $F(\omega)$ is the spectral distribution function defined in equations (1.17), (1.18). (Note an immediate deduction from (3): $F(\omega)$ is monotone non-decreasing.)

There are reasons (see Ex. (4.2.1)) why the Fourier representation should be the natural one, and have these properties. We shall merely note now that there are at least two useful properties: a statistical one, the orthogonality of $\zeta(\omega)$, and a transformation one, the possibility of relations such as (2.17), (2.25). We discuss this latter aspect in section (4).

The inverse of relations (1.17) and (1.18) holds:

$$\mathbf{\Gamma}_s = \frac{1}{2\pi}\int e^{i\omega s}\, d\mathbf{F}(\omega) \tag{4}$$

the limits of integration again being $\pm\pi$ or $\pm\infty$ in the two cases. Relation (4) is the *spectral representation of the covariance function.*

2.4 Linear Operations on Stationary Processes

Suppose the sequence x_t is *stationarily related* to the wide sense stationary sequence u_t by the m.a. formula

$$x_t = \sum_{j=-\infty}^{\infty} b_j u_{t-j} \tag{1}$$

Define

$$B(z) = \sum b_j z^j = \sum b_j e^{-ij\omega} \tag{2}$$

If

$$\int |B(z)|^2\, dF_{uu}(\omega) < \infty \tag{3}$$

then expression (1) exists as a mean square limit and the process $\{x_t\}$ is also stationary in the wide sense, and, in an obvious notation,

$$d\zeta_x(\omega) = B(z)\, d\zeta_u(\omega) \tag{4}$$

$$dF_{xx}(\omega) = |B(z)|^2\, dF_{uu}(\omega) \tag{5}$$

In particular, if the s.d.f. $f_{uu}(\omega)$ exists, then so does $f_{xx}(\omega)$, and

$$f_{xx}(\omega) = |B(z)|^2 f_{uu}(\omega) \tag{6}$$

Consider now the system (2.1), in which the u's are given in terms of the x's, rather than conversely. If t represents physical time, then we expect the $A(U)$ of (2.2) to be a one-sided operator

$$A(U) = \sum_0^{\infty} a_j U^j \tag{7}$$

and we expect to be able to invert to the one-sided m.a. relation (2.8). Formulae (4) – (6) of this section will be valid, with a one-sided operator, $B(U) = \sum_0^\infty b_j U^j$, determined by (2.6), if $A(z)$ is analytic in $|z| < 1$, and

$$\int |A(z)|^{-2}\, dF_{uu}(\omega) < \infty \tag{8}$$

If one does not have the restriction to one-sided operators (as would be the case in processes which were spatial rather than temporal), then functions such as $A(z)$ would be less restricted, and the appropriate inversions and representations determined by regularity conditions rather than by the idea of a physical distinction between past and future (cf. Whittle, 1954a).

The conditions and results of this section have obvious analogues in continuous time, with the substitution of $A(\omega)$ (as defined by (2.20)) for $A(z)$, etc.

2.5 The Elementary Stationary Processes

In section (2) we saw that there were two particular inputs u_t that were of interest: the pulse and the sinusoidal inputs. These might be called the *elementary inputs* in the time and frequency domains respectively.

If one demands that u_t be a wide sense stationary process, in discrete time, say, then there are two processes which stand in fairly immediate correspondence to these two elementary inputs. The elementary process in the time domain is a sequence of mutually uncorrelated variates, $\{\varepsilon_t\}$, each with zero mean and variance σ^2, say. For this process

$$\Gamma_s = \begin{Bmatrix} \sigma^2, & (s = 0) \\ 0 & (s \neq 0) \end{Bmatrix} \tag{1}$$

$$f_{\varepsilon\varepsilon}(\omega) = \sigma^2 \tag{2}$$

The elementary process in the frequency domain is again a sinusoid

$$u_t = \xi e^{i(\lambda t + \theta)} \tag{3}$$

but in which the amplitude ξ and the phase θ may be random variables, taking new values for each realisation of the process. Restrictions must be placed upon the joint distribution of ξ and θ if the process is to be stationary. If it is to be stationary in the wide sense, then $\xi \cos \theta$ and $\xi \sin \theta$ must have zero means, equal variances ($\sigma^2/2$, say) and zero correlation. If it is to be completely stationary, then θ must be distributed rectangularly upon $(-\pi, \pi)$, independently of ξ. Under either of these conditions one has

$$E(u_t) = 0 \tag{4}$$

$$\Gamma_s = \sigma^2 e^{i\lambda s} \tag{5}$$

$$f_{uu}(\omega) = \sigma^2 H(\omega - \lambda) \tag{6}$$

where H is the unit step function.

If one defines a process by the m.a. relation

$$x_t = \sum b_j \varepsilon_{t-j} \tag{7}$$

then the condition (4.3) becomes simply

$$\sum b_j^2 < \infty \tag{8}$$

and from (4.6) one has

$$f_{xx}(\omega) = g_{xx}(z) = \sigma^2 B(z) B(z^{-1}) \tag{9}$$

The process $\{x_t\}$ is said to be a *m.a. process*; a *finite m.a.* if the sum (7) is finite.

Suppose one has instead an *autoregression*:

$$\sum_0^\infty a_j x_{t-j} = \varepsilon_t \tag{10}$$

where t is physical time, so that (10) is to be regarded as determining x_t in terms of $\varepsilon_t, x_{t-1}, x_{t-2}, \ldots$. Then (10) can be inverted to the one-sided m.a. with $B(z) = A(z)^{-1}$, and

$$g_{xx}(z) = \frac{\sigma^2}{A(z)A(z^{-1})} \tag{11}$$

if $A(z)^{-1}$ is analytic in $|z| < 1$, and the coefficients in its expansion satisfy (8) (or, of course, $\int |A(z)|^{-2}\, d\omega < \infty$). If the summation in (10) extends up to a finite value, p, then the process is said to be a finite autoregression (a.r.) of order p, and the condition that (10) generate a stationary process then amounts simply to the demand that the polynomial $A(z)$ have no zeros in $|z| \leqslant 1$.

One can define a process of mixed type:

$$\sum_0^p c_j x_{t-j} = \sum_0^q d_j \varepsilon_{t-j} \tag{12}$$

for which

$$g_{xx}(z) = \sigma^2 \frac{D(z)D(z^{-1})}{C(z)C(z^{-1})} \tag{13}$$

in an obvious notation. The condition for validity is that $C(z)$ have no zeros in $|z| \leqslant 1$. This is an important type, a *process with rational s.d.f.* If, conversely, $g_{xx}(z)$ is rational, then x_t obeys a relation of type (12).

In continuous time there is a complication. Process (3) is still valid as the elementary process in the frequency domain (provided that θ is distributed uniformly over all real values, in the case of complete stationarity), but the concept of the uncorrelated process $\{\varepsilon_t\}$ cannot be carried over immediately. The difficulty is essentially that, as the auto-correlation of a stationary process becomes weaker, the variance of the variable, and the variance of its integral over a finite time interval, have different orders of magnitude. The way out usually adopted is to work with what might be regarded as the indefinite integral of an uncorrelated process: a process which is orthogonal, or of uncorrelated increments (just as one method of regularising the delta-function is to work with its "integral", the unit step-function). So, the continuous time analogue of the m.a. (7) will be

$$x_t = \int b_s \, d\eta_{t-s} \tag{14}$$

where $\{\eta_t\}$ is an orthogonal process for which

$$E(\eta_t) = 0 \tag{15}$$

$$E|\eta_t - \eta_s|^2 = \sigma^2(t - s) \qquad (s < t) \tag{16}$$

One has then

$$f_{xx}(\omega) = \sigma^2 B(\omega) B(-\omega) \tag{17}$$

if this expression is integrable.

One could construct various analogues of the a.r. If the a.r. is regarded as a stochastic difference equation, then an appropriate analogue would be a stochastic differential equation:

$$P(D)\int_\tau^t x_s \, ds = \eta_t - \eta_\tau \tag{18}$$

The equation must be written in this integrated form rather than as

$$P(D)x_t = \varepsilon_t \tag{19}$$

simply because of the fact that an uncorrelated input can be written only in integrated form. If P has all its roots in the left half-plane, then (18) generates a stationary process in time for which

$$f_{xx}(\omega) = \frac{\sigma^2}{P(i\omega)P(-i\omega)} \tag{20}$$

Despite the fact that an uncorrelated process may not exist, one can construct arbitrarily good approximations to it: for example, the process for which

$$f(\omega) = \frac{\sigma^2}{1 + \omega^2/\alpha^2} \tag{21}$$

$$\Gamma_s = \frac{\sigma^2 \alpha}{2} e^{-\alpha|s|} \tag{22}$$

where α is a large positive quantity. The fact that one can do this, and the convenience of being able to work with relations such as (19) rather than (18), makes it reasonable to postulate and work with an "uncorrelated process" ε_t (just as one finds it convenient to work with a "delta-function", knowing this to be a short-hand description of the limit of a sequence of functions). This idealised limit process would have constant s.d.f. and delta-function auto-correlation, as we see from (21), (22). It is spoken of by engineers as "white noise", which is a bad term; being incongruous as a mixed metaphor, and scientifically inaccurate (white light has not a uniform frequency spectrum, even over the visible range). We shall speak rather of "pure noise".

2.6 Linear Determinism: Wold's Decomposition

A deterministic process in discrete time would be one generated by a functional relation of the type (1.2.1), in which there is no random element. From now on we shall be almost entirely concerned with linear relations, and so we introduce the idea of *linear determinism.* A stationary process $\{x_t\}$ is *linearly deterministic* if x_t can be predicted by a linear function of $x_{t-1}, x_{t-2}, \ldots$ with zero mean square error. More specifically, if

$$\mathrm{v}_n = \min_{a_j} |x_t + \sum_1^n a_j x_{t-j}|^2 \quad (1)$$

then $\{x_t\}$ is linearly deterministic (or, henceforth, simply *deterministic*) if

$$\lim_{n \to \infty} \mathrm{v}_n = 0 \quad (2)$$

By considering linear predictors of successively increasing order n, as in (1), Wold (1938) was able to prove a remarkable theorem; that any stationary process $\{x_t\}$ can be uniquely represented as the sum of two mutually uncorrelated processes

$$x_t = \chi_t + \eta_t \quad (3)$$

where $\{\chi_t\}$ is deterministic, and $\{\eta_t\}$ can be represented as the one-sided moving average of a stationary uncorrelated sequence:

$$\eta_t = \sum_0^\infty b_s \varepsilon_{t-s} \quad (4)$$

The η_t component is referred to as *purely non-deterministic*; it has a limiting prediction error equal to $\sigma^2(\varepsilon)$. The extreme form of this type is the uncorrelated sequence $\{\varepsilon_t\}$ itself. The whole course of the χ_t process is determined by a set of initial values $(\ldots \chi_{\tau-1}, \chi_\tau)$, no matter how early τ is taken on the time axis. The process (5.3) represents an extreme, and elementary, type of a deterministic process.

The decomposition is a linear one (even though the process may be generated non-linearly) and the assertions concern second moments. The second moments of the process determine the decomposition.

A process may be deterministic in the strict sense, and yet non-deterministic (even purely non-deterministic) in the linear sense. For example (Doob, 1952), suppose that

$$x_t = e^{i(\lambda t + \theta)} \quad (5)$$

where λ is now a random variable, distributed independently of θ with probability distribution function $\frac{1}{2\pi}F(\lambda)$. Then

$$E(x_t) = 0$$

$$\Gamma_s = E(x_t \bar{x}_{t-s}) = E(e^{i\lambda s}) = \frac{1}{2\pi}\int e^{i\omega s}\, dF(\omega) \tag{6}$$

Comparison of (6) with (3.4) shows that $F(\omega)$ is actually the spectral distribution function of the process. One can choose $F(\omega)$ as the spectral distribution function of a purely non-deterministic process, and x_t would then itself be purely non-deterministic, despite its simple functional form (5). The point is that, given a series of observations $(x_s; s \leqslant t)$, one would have to operate on this non-linearly if one wished to calculate the value of λ holding for the particular realisation. Having done this, one could predict without error, but the prediction would be a non-linear one.

A rather different example is the following, due to Moran: suppose x_0 is randomly and uniformly distributed in (0, 1), and that

$$x_t = \text{fractional part of } (2x_{t-1}) \qquad (t = 1, 2, \ldots) \tag{7}$$

The process is thus deterministic in the strict sense. Now, if x_0 has the binary representation

$$x_0 = 0\,.\,\nu_0\nu_1\nu_2, \ldots \tag{8}$$

then the ν_j are independent random variables taking the values 0, 1 each with probability $\frac{1}{2}$, and

$$x_t = 0\,.\,\nu_t\nu_{t+1}\nu_{t+2}\ldots \tag{9}$$

We thus find

$$E(x_t) = \tfrac{1}{2} \tag{10}$$

$$\Gamma_s = \operatorname{cov}(x_t, x_{t-s}) = \frac{2^{-s}}{12} \qquad (t, t-s \geqslant 0) \tag{11}$$

From this it can be shown (see section (3.3)) that the best linear predictor of x_t in terms of preceding values is

$$\hat{x}_t = \tfrac{1}{4} + \tfrac{1}{2}x_{t-1} \tag{12}$$

and that this has m.s.e. $\frac{1}{16}$.

2.7 Moving Average and Autoregressive Representations of a Process

Let us return to the spectral representations (3.1) and (3.4). The function $F(\omega)$ is monotone non-decreasing, and so can be represented as the sum of three such functions, each with special properties

$$F(\omega) = F_d(\omega) + F_c(\omega) + F_s(\omega) \tag{1}$$

Here $F_d(\omega)$ is a purely discontinuous function, increasing only on a denumerable set of ω values, and so consisting of denumerably many positive "steps". The function $F_c(\omega)$ is absolutely continuous, and so is the integral of its derivative (which will be positive). The singular component, $F_s(\omega)$, is continuous, but has a derivative equal either to 0 or $+\infty$.

Cramer (1942) showed that, corresponding to this decomposition, there is a decomposition of the variate:

$$x_t = x_{dt} + x_{ct} + x_{st} \tag{2}$$

the three processes $\{x_{dt}\}$, $\{x_{ct}\}$ and $\{x_{st}\}$ being mutually uncorrelated and having spectral distribution functions F_d, F_c and F_s respectively. The x_{dt} process can be thought of as a sum of terms such as (5.3), or the limit of such a sum. It is deterministic; having observed x_{dt} up to a certain point in time, one can construct linear predictors which are free from error. The singular component, x_{st}, is also deterministic. It is not easily characterised, and does not naturally occur in temporal processes. We shall not be further concerned with it. The absolutely continuous component, x_{ct}, is important: the m.a. and a.r. processes are of this type. It constitutes the essentially statistical component of the process.

A m.a. process (i.e. a process specified by (5.7), (5.8)), has absolutely continuous spectrum, and, conversely, a process with absolutely continuous spectrum can be represented as a m.a. For, in this latter case the process has an integrable s.d.f. $f(\omega)$, which, being positive, can be represented

$$f(\omega) = |\phi(\omega)|^2 \tag{3}$$

and the spectral representation of the variate can be written

$$x_t = \frac{1}{\sqrt{2\pi}} \int e^{i\omega t} \phi(\omega)\, d\zeta^*(\omega) \tag{4}$$

where

$$\zeta^*(\omega) = \int^{\omega} \frac{d\zeta(\omega)}{\phi(\omega)} \tag{5}$$

so that ζ^* is also an orthogonal random function for which

$$E|d\zeta^*(\omega)|^2 = \frac{dF(\omega)}{|\phi(\omega)|^2} = d\omega \tag{6}$$

Now, we know from (3) that $\phi(\omega)$ is square integrable, and so has a Fourier expansion

$$\phi(\omega) = \sum_j b_j e^{-i\omega j} \tag{7}$$

which represents it, at least in the mean. Relation (4) can thus be written in the form (5.7) if we define

$$\varepsilon_t = \frac{1}{\sqrt{2\pi}} \int e^{i\omega t}\, d\zeta^*(\omega) \tag{8}$$

and it follows immediately from (6) that the ε_t constitute a stationary uncorrelated sequence. The process is thus representable as a m.a.

In fact, it is evident that there are infinitely many functions $\phi(\omega)$ satisfying (3), and so infinitely many m.a. representations. However, for a process in time there is only one physically meaningful m.a. representation: that involving no "future" ε values, so that in (5.7) b_j is zero if j is negative. The question of whether such a one-sided m.a. exists is related to that of linear determinism. The criterion is the following: the process is purely non-deterministic (and so has a one-sided m.a. representation) if and only if

$$\int_{-\pi}^{\pi} \log f(\omega)\, d\omega > -\infty \tag{9}$$

Otherwise it is deterministic. The corresponding criterion for continuous time processes is

$$\int_{-\infty}^{\infty} \frac{\log f(\omega)\, d\omega}{1 + \omega^2} > -\infty \tag{10}$$

(For proofs see Doob, 1953.)

2.8 The Canonical Factorisation of the Spectral Density Function

It is apparent that the finding of a one-sided m.a. representation is equivalent (in the discrete case) to the finding of a representation (7.3) in which $\phi(\omega)$ has a Fourier expansion containing no negative powers of $z = e^{-i\omega}$. Suppose this exists, and is written

$$f(\omega) = g(z) = \sigma^2 B(z) B(z^{-1}) \tag{1}$$

where we have taken out a factor σ^2 so as to achieve the standardisation $b_0 = 1$. Now, suppose that $\log g(z)$ has a Laurent expansion valid in some annulus $\rho < |z| < \rho^{-1}$, ($\rho < 1$), so that

$$g(z) = \exp\left[\sum_{-\infty}^{\infty} c_j z^j\right] \tag{2}$$

in this region. Then

$$B(z) = \exp\left[\sum_{1}^{\infty} c_j z^j\right] \tag{3}$$

is evidently an appropriate choice, for this will be analytic in $|z| < \rho^{-1}$, and so have a Taylor expansion in powers of z. The m.a. representation exists, with

$$\sigma^2(\varepsilon) = e^{c_0} = \exp\left[\frac{1}{2\pi}\int_{-\pi}^{\pi} \log f(\omega)\, d\omega\right] \tag{4}$$

the so-called *prediction variance*. (The motivation for the term will become apparent in section (3.2).)

In fact, there is also an *autoregressive representation*, for it follows from (3) that $B(z)^{-1}$ is also analytic in $|z| < \rho^{-1}$, and so can be expanded in this region

$$B(z)^{-1} = \sum_0^\infty a_j z^j \tag{5}$$

The m.a. representation

$$x_t = \sum_0^\infty b_j \varepsilon_{t-j} \tag{6}$$

can thus be inverted to

$$\sum_0^\infty a_j x_{t-j} = \varepsilon_t \tag{7}$$

for the sums in (6) and (7) will both be m.s. convergent.

The condition (7.9) is much weaker than the demand of analyticity, and leads to rather weaker conclusions. Relations (2) and (3) are now valid as Fourier representations for real ω, i.e. on $|z| = 1$ alone, so that there is a representation (6). However, one cannot infer analyticity, in however narrow an annulus, so it cannot in general be concluded that an a.r. representation (7) can be found, except possibly in some limiting sense. Suppose, for example, that

$$f(\omega) = \sigma^2(1 - z)(1 - z^{-1}) = 4\sigma^2 \sin^2\left(\frac{\omega}{2}\right) \tag{8}$$

Then $\log f(\omega)$ has a Fourier expansion which represents it in the mean:

$$\log f(\omega) = \log \sigma^2 - \sum_1^\infty \frac{e^{i\omega j} + e^{-i\omega j}}{j} \tag{9}$$

whence one derives the canonical factor and m.a. representation

$$B(z) = 1 - z \tag{10}$$

$$x_t = \varepsilon_t - \varepsilon_{t-1} \tag{11}$$

However, (11) can evidently not be inverted to a.r. relation as it stands (see section (3.8)). This is due to the fact that $f(\omega)$ has a zero at $\omega = 0$, which is serious enough to prevent $f(\omega)^{-1}$ having a Fourier representation.

The canonical factorisation (1) is extremely important for prediction theory. Equation (3) evidently supplies an explicit method of calculating the canonical factor, when it exists.

The representation (1) is not unique unless one sets conditions on the zeros of $B(z)$, as well as on its region of analyticity; cf. the example

$$\begin{aligned} g(z) &= (1 - \beta z)(1 - \beta z^{-1}) \\ &= \beta^2(1 - \beta^{-1}z)(1 - \beta^{-1}z^{-1}) \end{aligned} \tag{12}$$

If one makes the demand that $B(z)$ be analytic in $|z| \leqslant 1$, $B(z)^{-1}$ in $|z| < 1$, then the factorisation is unique, and is that determined by (3), provided $g(z)$ has no zeros on $|z| = 1$.

For the particular case of *rational s.d.f.'s* there is a simpler way of calculating $B(z)$. Suppose $g(z)$ can be written as a ratio of polynomials

$$g(z) = \frac{P(z)}{Q(z)} \tag{13}$$

We know that $g(z)$ has a symmetric Laurent expansion on $|z| = 1$, so that if $z = \zeta$ is a pole (or zero), then so is $z = \zeta^{-1}$. We can thus write

$$g(z) = z^{\mu}\sigma^2 \frac{\prod_j (1 - \beta_j z)(1 - \beta_j z^{-1})}{\prod_k (1 - \alpha_k z)(1 - \alpha_k z^{-1})} \tag{14}$$

where $|\beta_j| \leqslant 1$, $|\alpha_k| < 1$, all j, k. On the same grounds of symmetry, the exponent μ must be zero. We have then, evidently,

$$B(z) = \frac{\prod(1 - \beta_j z)}{\prod(1 - \alpha_k z)} \tag{15}$$

while the σ^2 of (14) can be identified with that of (1).

There are analogous results for processes in continuous time. If condition (7.10) is satisfied, then one can find a representation

$$f(\omega) = B(\omega)B(-\omega) \tag{16}$$

where $B(\omega)$ is analytic in $Im(\omega) \leqslant 0$, and $B(\omega)^{-1}$ in $Im(\omega) < 0$. Since $f(\omega)$ has a symmetric Fourier expansion, a rational s.d.f. can always be written

$$f(\omega) = \kappa^2 \frac{\prod(\beta_j + i\omega)(\beta_j - i\omega)}{\prod(\alpha_k + i\omega)(\alpha_k - i\omega)} \tag{17}$$

where $Re(\beta_j) \geqslant 0$, $Re(\alpha_k) > 0$, all j, k, so that in this case

$$B(\omega) = \kappa \frac{\prod(\beta_j + i\omega)}{\prod(\alpha_k + i\omega)} \tag{18}$$

We shall have occasion to factorise functions other than s.d.f.'s, so the following convention is useful: that if it is said that a function h has the canonical factorisation

$$h = |P|^2$$

then it will be understood that $P(z)$ (or $P(\omega)$, in continuous time) is the factor for which both P and P^{-1} are analytic in $|z| \leqslant 1$ (or $Im(\omega) \leqslant 0$, in continuous time). The statement will also imply the assumption that h can be so factorised.

2.9 Multivariate Processes

Most of the results of sections (2) to (8) generalise fairly directly to the case where the process variate $\mathbf{x}_t$ is an $r \times 1$ vector, if the notation is generalised as in section (1). Thus, the spectral representations (3.1) and (3.4) hold, with (3.3) replaced by

$$E[\boldsymbol{\zeta}(\omega_2) - \boldsymbol{\zeta}(\omega_1)][\boldsymbol{\zeta}(\omega_2) - \boldsymbol{\zeta}(\omega_1)]^{\dagger} = \mathbf{F}(\omega_2) - \mathbf{F}(\omega_1) \quad (\omega_1 < \omega_2) \tag{1}$$

and (3.2) similarly modified.

The spectral density matrix (s.d.m.) $\mathbf{F}(\omega)$ is Hermitian, and its increments, $\mathbf{F}(\omega_2) - \mathbf{F}(\omega_1)$, (or $\mathbf{dF}(\omega)$) are always non-negative definite as matrices. This implies, in particular, that if x and y are scalar components of a multivariate stationary process, then

$$|dF_{xy}|^2 = dF_{xy}dF_{yx} \leqslant dF_{xx}dF_{yy} \tag{2}$$

so that if dF_{xx} or dF_{yy} is zero, then so is dF_{xy}.

If

$$\mathbf{x}_t = \sum \mathbf{b}_j\, \varepsilon_{t-j} \tag{3}$$

$$\mathbf{B}(z) = \sum \mathbf{b}_j z^j = \sum \mathbf{b}_j e^{-i\omega j} \tag{4}$$

where the coefficients $\mathbf{b}_j$ are $r \times r$ matrices, then expression is valid as a m.s. limit, and defines a stationary process with finite variances, if

$$\int_{-\pi}^{\pi} \mathbf{B}(z) d\mathbf{F}_{\varepsilon\varepsilon}(\omega)\mathbf{B}(z)^{\dagger} < \infty \tag{5}$$

If (5) is fulfilled, then

$$d\mathbf{F}_{\mathbf{xx}}(\omega) = \mathbf{B}(z)d\mathbf{F}_{\varepsilon\varepsilon}(\omega)\mathbf{B}(z)^{\dagger} \tag{6}$$

$$d\mathbf{F}_{\mathbf{xy}}(\omega) = \mathbf{B}(z)d\mathbf{F}_{\varepsilon\mathbf{y}}(\omega) \tag{7}$$

$$d\mathbf{F}_{\mathbf{xx}}(\omega) = d\mathbf{F}_{\mathbf{y}\varepsilon}(\omega)\mathbf{B}(z)^{\dagger} \tag{8}$$

In particular, if the $\boldsymbol{\epsilon}_t$ in (3) constitute a stationary uncorrelated sequence, so that

$$\mathbf{\Gamma}_s{}^{(\epsilon)} = E(\varepsilon_t \varepsilon'_{t-s}) = \mathbf{V}\delta_s \tag{9}$$

say, then (3) is a m.a. model in the more special sense, and

$$\mathbf{f}_{\mathbf{xx}}(\omega) = \mathbf{B}(z)\mathbf{V}\mathbf{B}(z)^{\dagger} \tag{10}$$

Similarly, for the finite multivariate a.r.

$$\sum_0^p \mathbf{a}_j \mathbf{x}_{t-j} = \varepsilon_t \tag{11}$$

one has

$$\mathbf{f}(\omega) = \mathbf{A}(z)^{-1}\mathbf{V}[\mathbf{A}(z)^{\dagger}]^{-1} \tag{12}$$

The condition for the a.r. to generate a stationary process, if (11) is understood as a recursive relation in time, is that all the roots of

$$|\mathbf{A}(z)| \equiv |\sum_0^p \mathbf{a}_j z^j| = 0 \tag{13}$$

should lie outside the unit circle.

The multivariate a.r., in particular that for $p = 1$, is an important model. One feels that if one has included all relevant variables in one's model, then the process should be Markov; if one adds the assumption of linearity, then one is led to a first-order multivariate a.r. The s.d.f. of any component scalar variate will be rational, and this is one justification for the common restriction to rational s.d.f.'s. For example, the s.d.f. $f_{11}(\omega)$ of the component process $\{x_{1t}\}$ will be the (11) element in the matrix (12). If $\mathbf{a}_0$, $\mathbf{a}_p$ and $\mathbf{V}$ are all non-singular, then this will be of the form

$$f_{11}(\omega) = \left|\frac{C(z)}{D(z)}\right|^2 \tag{14}$$

where $D(z)$ is a polynomial of degree rp exactly, and $C(z)$ is a polynomial of degree at most $(r - 1)p$.

CHAPTER 3

A FIRST SOLUTION OF THE PREDICTION PROBLEM

In this chapter we consider a purely non-deterministic process $\{x_t\}$, and construct the l.l.s.e. of a variate y which is based upon the sample of past values $(x_s;\ s \leqslant t)$. An important case is that in which y is a term y_t in a process $\{y_t\}$ jointly stationary with $\{x_t\}$. If $y_t = x_{t+\nu}$, then the problem is one of *pure prediction*, and most of our examples concern this special case. More general examples will be given in Chapter 6, where we describe Wiener's method. The method used in this chapter is essentially a specialised version of one due to Kolmogorov [1939, 1941], which in turn originated in work of Wold's [1938].

Pre-requisites: Chapter 2.

3.1. Deterministic Components

We shall restrict attention to the purely deterministic process, because deterministic components can always be subtracted out, and in most practical cases their prediction is trivial. In fact, only rarely will one have anything more sophisticated than a trigonometric sum,

$$\chi_t = \sum_j \beta_j \exp [i\omega_j t]$$

This is the sort of term one would expect in tidal prediction, for example.

The spectral distribution function, supposed known, gives one information in the frequencies ω_j and absolute amplitudes $|\beta_j|$. Again, in a practical case, if one has as much information as this, one probably also knows the phases, and so the actual β_j. In any event, these can be estimated from a sample by the methods described in sections (4.3) and (8.3).

3.2 Pure Prediction

Suppose that the spectral density function $f(\omega) = g(z)$ of $\{x_t\}$ is analytic in an annulus $\rho < |z| < \rho^{-1}$, where $0 \leqslant \rho < 1$, so that the process has both moving average and autoregressive representations.

Then the l.l.s. predictor of $x_{t+\nu}$ can be represented in either of the two forms

$$\hat{x}_{t+\nu} = \sum_0^\infty \gamma_j x_{t-j} = \sum_0^\infty \phi_j \varepsilon_{t-j} \tag{1}$$

and the generating functions of the two sets of coefficients

$$\gamma(z) = \sum_0^\infty \gamma_j z^j$$
$$\phi(z) = \sum_0^\infty \phi_j z^j \tag{2}$$

will be connected by the relation

$$\phi(z) = B(z)\gamma(z) \tag{3}$$

We can be sure that these generating functions converge at least inside the unit circle, because $\hat{x}_{t+\nu}$ will certainly have finite variance, so that

$$\sum \phi_j^2 < \infty \tag{4}$$

Now

$$\begin{aligned}\sigma_\nu^2 &= E[\hat{x}_{t+\nu} - x_{t+\nu}]^2 \\ &= E[\sum_0^\infty b_j \varepsilon_{t+\nu-j} - \sum_0^\infty \phi_j \varepsilon_{t-j}]^2 \\ &= \sigma^2[\sum_0^{\nu-1} b_j^2 + \sum_0^\infty (\phi_j - b_{j+\nu})^2]\end{aligned} \tag{5}$$

It is clear that this expression, regarded as a function of $\phi_0, \phi_1, \ldots$, is least when

$$\phi_j = b_{j+\nu} \qquad (j \geqslant 0) \tag{6}$$

and with this the second form of the l.l.s. predictor (1) is determined. One takes the moving average representation of $x_{t+\nu}$, and sets the unknown future disturbances $\varepsilon_{t+1}, \varepsilon_{t+2}, \ldots \varepsilon_{t+\nu}$ equal to their expected values; zero.

It is, however, the first form of the predictor (1) which is the more convenient, since this expresses $\hat{x}_{t+\nu}$ directly in terms of observed quantities; $x_t, x_{t-1}, \ldots$ We have, in virtue of (3) and (6),

$$\gamma(z) = \frac{\sum_{j=0}^\infty b_{j+\nu} z^j}{B(z)} = \frac{1}{B(z)}\left[\frac{B(z)}{z^\nu}\right]_+ \tag{7}$$

This is the basic formula, determining the coefficients of the l.l.s. predictor. From (5) and (6) we see that the prediction variance is

$$\sigma_\nu^2 = \sigma^2 \sum_0^{\nu-1} b_j^2 \tag{8}$$

3.3 Some Examples

Consider the prediction of the autoregressive scheme

$$x_t - \alpha x_{t-1} = \varepsilon_t \tag{1}$$

for which

$$B(z) = (1 - \alpha z)^{-1} \tag{2}$$

and (2.7) becomes

$$\gamma(z) = (1 - \alpha z) \sum_0^\infty \alpha^{j+\nu} z^j = \alpha^\nu \tag{3}$$

The predictor and its mean square error are simply

$$\hat{x}_{t+\nu} = \alpha^\nu x_t \tag{4}$$

$$\sigma_\nu^2 = \sigma^2 \sum_0^{\nu-1} \alpha^{2j} = \frac{\sigma^2(1 - \alpha^{2\nu})}{1 - \alpha^2} \tag{5}$$

As ν increases $\hat{x}_{t+\nu}$ tends to the unconditioned value of x_t, zero, and σ_ν^2 tends to the unconditioned variance, $\sigma^2/(1 - \alpha^2)$.

In general, it will be true for an autoregression of order p that $\hat{x}_{t+\nu}$ is a function of the last p observed x values alone, and is in fact recursively determined by the autoregressive relation itself

$$\sum_0^p a_s x_{t-s} = \varepsilon_\tau \qquad (\tau = t + 1, t + 2 \ldots t + \nu) \tag{6}$$

with the values ε_τ set equal to zero. This is true, because $x_{t+\nu}$ is determined from the relations (6), and we know from (2.6) that the same relations must hold for $\hat{x}_{t+\nu}$ with the substitution

$$\varepsilon_{t+1} = \ldots = \varepsilon_{t+\nu} = 0.$$

Ex. 1. Verify directly from equation (2.7) that for $\nu = 1$ we can write

$$\hat{x}_{t+1} = - \sum_1^\infty a_j x_{t-j} \tag{7}$$

Ex. 2. Show from (2.7) that if $x_s = y^{-s}$ $(s \leqslant t)$ then

$$\sum_{\nu=1}^\infty \hat{x}_{t+\nu} z^{t+\nu} = -\frac{y^{-t} z^{t+1}}{A(z)} \frac{A(z) - A(y)}{z - y}, \quad (|y|, |z| < \rho^{-1}) \tag{8}$$

so that the l.l.s. predictor agrees with the solution (2.2.14) to the deterministic "initial-value problem". This is, of course, no more than has just been shown, but relation (8) has its own interest.

Ex. 3. Consider a second-order autoregression with $A(z) = (1 - \alpha_1 z)(1 - \alpha_2 z)$. Show that

$$\gamma(z) = \frac{\alpha_1^{\nu+1}(1 - \alpha_2 z) - \alpha_2^{\nu+1}(1 - \alpha_1 z)}{\alpha_1 - \alpha_2}$$

and calculate the corresponding σ_ν^2.

Consider the first-order moving-average scheme

$$x_t = \varepsilon_t - \beta \varepsilon_{t-1} \tag{9}$$

If $|\beta| < 1$ then $B(z) = 1 - \beta z$, and we can apply (2.7) to obtain

$$\left.\begin{aligned} \gamma(z) &= -\frac{\beta}{1 - \beta z} & (\nu = 1) \\ \gamma(z) &= 0 & (\nu > 1) \end{aligned}\right] \tag{10}$$

Thus if $\nu = 1$ the predictor will be

$$\hat{x}_{t+1} = - \sum_0^\infty \beta^{j+1} x_{t-j} \tag{11}$$

(again, just the autoregressive representation with a zero residual) and $\sigma_1^2 = \sigma^2$. If $\nu > 1$, then the observed past conveys no information on $x_{t+\nu}$, so that the best estimate is just the unconditioned mean value, and σ^2 equals the unconditioned x variance, $\sigma^2(1 + \beta^2)$. In general, it is plain from (2.7) and (2.8) that, for a moving average of order q, $\gamma(z)$ will be zero and σ_ν^2 will have its full value if $\nu > q$.

If $|\beta| > 1$ in relation (9), then (2.6) will still be correct, but the inversion to an autoregression implied in (2.7) will be invalid. In order to use (2.7) one will have to calculate the canonical form of $B(z)$:

$$\begin{aligned} g(z) &= \sigma^2(\varepsilon)(1 - \beta z)(1 - \beta z^{-1}) \\ &= \beta^2\sigma^2(\varepsilon)(1 - \beta^{-1}z)(1 - \beta^{-1}z^{-1}) \end{aligned} \tag{12}$$

That is, in the canonical moving average $B(z) = 1 - \beta^{-1}z$, and the variance of the residual is $\sigma^2 = \beta^2\sigma^2(\varepsilon)$, $\sigma^2(\varepsilon)$ being the variance of ε in the original relation (9).

If $|\beta| = 1$, then our treatment seems to fail: the process then has no obviously valid autoregressive representation. We shall consider this point in section (8).

Ex. 4. If $B(z) = (1 - \beta_1 z)(1 - \beta_2 z)$ show that

$$\hat{x}_{t+1} = -\sum_0^\infty(\Delta_{j+2}/\Delta_1)x_{t-j}, \quad \sigma_1^2 = \sigma^2,$$

$$\hat{x}_{t+2} = \beta_1\beta_2\sum_0^\infty(\Delta_{j+1}/\Delta_1)x_{t-j}. \ \sigma_2^2 = \sigma^2[1 + (\beta_1 + \beta_2)^2],$$

where $\Delta_j = \beta_1^j - \beta_2^j$.

Both the moving average and autoregressive processes are included in the general class of *processes with rational s.d.f.* For such processes, both $B(z)$ and $\gamma(z)$ will also be rational, and may be determined relatively easily (provided always that $g(z)$ has no zeros *on* the unit circle).

Ex. 5. Suppose that in the canonical moving-average form

$$B(z) = \frac{Q_0(z)}{P(z)}$$

where Q_0 and P are polynomials, and that repeated division yields

$$B(z) = \sum_0^{\nu-1} b_j z^j + \frac{z^\nu Q_\nu(z)}{P(z)}$$

where Q_ν is also a polynomial. Show that for prediction ν steps ahead

$$\gamma(z) = \frac{Q_\nu(z)}{Q_0(z)}.$$

Ex. 6. If $B(z) = \dfrac{1 - \beta z}{1 - \alpha z}$ show that

$$\hat{x}_{t+\nu} = (\alpha - \beta)\alpha^{\nu-1}\sum_0^\infty \beta^j x_{t-j}$$

$$\sigma_\nu^2 = \sigma^2\left[1 + \frac{(\alpha - \beta)^2(1 - \alpha^{2\nu-2})}{1 - \alpha^2}\right].$$

Ex. 7. Suppose x_t is an first-order autoregressive variable with superimposed error, so that

$$x_t = u_t + \varepsilon_t'$$
$$u_t = \alpha u_{t-1} + \varepsilon_t$$

where $\{\varepsilon_t\}$, $\{\varepsilon_t'\}$ are both internally and mutually uncorrelated, with variances σ_1^2, $\lambda\sigma_1^2$ respectively. Show that for pure prediction of x the calculations of Ex. 6 can be applied with

$$\beta = \frac{1 + \lambda(1 + \alpha^2) - \Delta}{2\lambda\alpha}$$
$$\Delta = \sqrt{1 + 2\lambda(1 + \alpha^2) + \lambda^2(1 - \alpha^2)^2}$$
$$\sigma^2 = \frac{\sigma_1^2}{(1 - \alpha\beta)(1 - \alpha^{-1}\beta)} = \frac{\lambda\alpha\sigma_1^2}{\beta}.$$

Ex. 8. If

$$B(z) = \frac{1 - \beta z}{(1 - \alpha_1 z)(1 - \alpha_2 z)}$$

show that

$$\gamma(z) = \frac{(\alpha_1 - \beta)\alpha_1^\nu(1 - \alpha_2 z) - (\alpha_2 - \beta)\alpha_2^\nu(1 - \alpha_1 z)}{(\alpha_1 - \alpha_2)(1 - \beta z)}$$

so that the predictor is of the form

$$\hat{x}_{t+\nu} = cx_t + d\sum_0^\infty \beta^j x_{t-j}.$$

Ex. 9. Show from the three equations of Ex. 5 that if $P(z)$ has a simple zero at $z = \alpha^{-1}$, then $\gamma(\alpha^{-1}) = \alpha^\nu$, so that the prediction relation is exact for a sequence $x_t = \alpha^t$.

Ex. 10. Show that if $(1 - \alpha z)$ is a k-fold factor of $P(z)$, then the sequences $x_t = t^j\alpha^t$, $(j = 0, 1 \ldots k - 1)$ are all predicted exactly.

Ex. 11. Show that if the polynomials $P(z)$ and $Q_0(z)$ are of degrees p and q, and $\nu \geqslant q - p + 1$, then the results of Exs. 9 and 10 can be used to determine $Q_\nu(z)$.

Ex. 12. Let $\hat{x}_{t+\nu}$ be denoted more exactly $\hat{x}_{t+\nu,\,\nu}$: the estimate of $x_{t+\nu}$ based on data an interval ν in the past. Suppose that, in the notation of Ex. 5, $Q_r(z) = \sum_j c_{rj}z^j$. Show then that

$$\sum_j c_{0j}\hat{x}_{t+\nu-j,\,\nu} = \sum_j c_{\nu j}x_{t-j}.$$

Thus, for the example of Ex. 6,

$$\hat{x}_{t+\nu,\,\nu} - \beta\hat{x}_{t+\nu-1,\,\nu} = (\alpha - \beta)\alpha^\nu x_t.$$

These relations are often convenient for the recursive calculation of forecasts as time advances, and new data become available.

3.4 The Factorisation of the Spectral Density Function in Practice

It will be clear from the last two sections that an essential preliminary to the calculation of the predictor is the determination of the canonical factorisation

$$g(z) = \sigma^2 B(z)B(z^{-1}) \tag{1}$$

In practice, $g(z)$ will seldom be known analytically: one will have to work from numerical values of the auto-covariances Γ_s or, possibly, of the s.d.f. $f(\omega) = g(e^{-i\omega})$ itself.

One method that has been suggested for achieving the factorisation is to approximate $g(z)$ by a finite sum

$$g(z) = \sum_{-q}^{q} \Gamma_s\, z^s \tag{2}$$

and determine the roots of this polynomial numerically. This is equivalent to approximating the process by a m.a. of order q.

However, most processes are much better fitted by a low-order a.r. than by a low-order m.a., for the simple reason that this is often nearer the true model. Under these circumstances it seems reasonable to attempt a representation

$$g(z) = \frac{\sigma^2}{A(z)A(z^{-1})} \tag{3}$$

where $A(z)$ is a low-order polynomial, of order p, say. If the true $A(z)$ were of order p, then we should have

$$\sum_{k=0}^{p} a_k \Gamma_{j-k} = \begin{bmatrix} \sigma^2 & (j=0) \\ 0 & (j>0) \end{bmatrix} \tag{4}$$

and we can use these relations for $j = 0, 1, 2, \ldots p$ to determine σ^2 and $a_1, a_2, \ldots a_p$ (a_0 being set equal to unity). If the scheme is of higher order than p, and it will in general be of infinite order, then the infinite equation system (4) (with $p = \infty$) will be required to determine $A(z)$. However, the finite equation system will determine a polynomial approximation to $A(z)$, which is best in the sense that it minimises $v_p = E|\sum a_k x_{t-k}|^2 = \sum\sum a_j a_k \Gamma_{j-k}$, and this is just the m.s. error for a pure prediction one step ahead.

Thus, suppose the a_k and σ^2 determined from the first $p + 1$ equations of (4) are denoted a_{pk}, v_p, and that $A_p(z) = \sum_0^p a_{pk} z^k$.

If this autoregressive operator is used to predict one step ahead:

$$\hat{x}_{t+1} = -\sum_1^p a_{pk} x_{t-k} \tag{5}$$

then the m.s. prediction error will be exactly v_p, even if $A_p(z)$ is not equal to $A(z)$, i.e. if p is not sufficiently large. On the other hand, it is *not* true that the predictor of $x_{t+\nu}$ determined by ν-fold application of (5) has then m.s. error appropriate to the a.r. characterised by $A_p(z)$, v_p, for $\nu > 1$, although if $A_p(z)$ is sufficiently near $A(z)$ the difference will be small.

Example. Suppose one fits a first-order a.r.:

$$x_t + a_{11} x_{t-1} = \varepsilon_t$$

with

$$\alpha_{11} = -\Gamma_1/\Gamma_0 = -\rho_1; \; v_1 = \Gamma_0(1 - \rho_1^2).$$

The approximate $\hat{x}_{t+\nu}$ obtained by use of this scheme is $\rho_1^\nu x_t$, with m.s. error $\Gamma_0(1 - 2\rho_\nu \rho_1^\nu + \rho_1^{2\nu})$, which for $\nu > 1$ may be either smaller or larger than $\Gamma_0(1 - \rho_1^{2\nu})$, the m.s.e. if the a.r.1 scheme were the true model.

If $A_p(z)$ is to approximate $A(z)$ satisfactorily in the canonical representation (3) it is essential that $A_p(z)$ should have all its zeros outside the unit circle, so that $A_p(z)^{-1}$ is analytic in $|z| \leqslant 1$. This is true, and is in fact a classic result in the theory of orthogonal polynomials.

What one will do in practice is to fit a.r.'s of successively increasing order, $p = 1, 2, 3, \ldots$ The value of v_p will diminish, tending monotonically to the limit value σ^2. One stops when the value of v_p has levelled out, and shows little likelihood of further substantial decrease ("likelihood", because one can never be sure).

It is unnecessary to solve a complete equation system for each new value of p: Durbin (1960) has described a useful recursive method which gives the values of the $a_{p+1,\,k}$ in terms of the a_{pk}. In fact, if

$$v_p = \sum_{k=0}^{p} a_{pk}\Gamma_{-k} \tag{7}$$

$$\Delta_p = \sum_{k=0}^{p} a_{pk}\Gamma_{p-k+1} \tag{8}$$

then

$$a_{p+1,k} = a_{pk} + a_{p+1,p+1}a_{p,p-k+1} \qquad (k = 1, 2, \ldots p) \tag{9}$$

$$a_{p+1,p+1} = -\Delta_p/v_p \tag{10}$$

The proof follows fairly directly from the fact that the equations for the determination of the $a_{p+1,\,k}$ can be written

$$\Gamma_j + a_{p+1,p+1}\,\Gamma_{-p+j-1} + \sum_{k=1}^{p} \Gamma_{j-k}a_{p+1,k} = 0 \qquad (j = 1, 2, \ldots p) \tag{11}$$

$$a_{p+1,p+1}\Gamma_0 + \sum_{k=0}^{p} a_{p+1,k}\Gamma_{p-k+1} = 0 \tag{12}$$

Equation (9) follows from (11); (10) from substitution of (9) into (12).

A numerical example may be of interest; we quote the following figures from an analysis of observations on water height in a small coastal channel (Whittle, 1954b). The object here was to fit a statistical model rather than to predict, but in doing this attempts were made to fit an a.r. on the above lines.

Fitting a.r.'s of orders $p = 0, 1, 2, \ldots$ directly one obtained the following table:

p . .	0	1	2	3	4	5	6
v_p . .	99·01	20·76	8·01	7·86	7·62	7·62	7·60

In fact, the calculation was continued, with scarcely any decrease in v_p. Judging from the above table, an a.r. of order 3, or even 2, predicts one step ahead with an m.s.e. which seems to be only fractionally above the likely limit value.

Despite this, however, the observed and fitted spectral densities differed considerably; the observed density showing much more "structure" in the way of peaks which could not be attributed to random variation. It was in fact shown statistically that the series contained two sinusoidal components. These were estimated and subtracted out; fitting of a.r.'s to the residual series gave the following table:

p . .	0	1	2	3	4	5	6
v_p . .	78·70	16·99	7·62	7·26	7·17	7·02	7·02

This model reproduced the features of the observed frequency spectrum very much more closely than the first one; the difference in fit was very significant statistically. Nevertheless, the actual difference in fit, as measured by v_p, is not large. Thus, for $p = 4$ we have $v = 7{\cdot}62$ and 7·17 respectively. This is interesting: that a model which is rather far from the truth has almost as good a predictive power (at least for $\nu = 1$) as a model which is very much nearer the truth. Of course, as pointed out after equation (5), the prediction errors of the two schemes for $\nu > 1$ may be very different.

The fact that two models may have very similar values of σ_1^2 (m.s.e. for prediction one step ahead) and yet differ very much in physical adequacy, and probably also in σ_ν^2 for $\nu > 1$, makes one realise the necessity for caution.

3.5 Pure Prediction in Continuous Time

The case of continuous time is formally analogous to that of discrete time, and we shall rarely give separate derivations. There are certainly difficulties of rigour; for example, the ideas of a pulse input (δ-function), a purely random sequence or the Fourier transform of the weight function corresponding to a difference operator ($1 - z = 1 - e^{-i\omega}$) are all relatively elementary and unexceptionable ideas in discrete time, but their analogues in continuous time all require special discussion. To some extent these are side-issues to our main theme, and we shall not consider them in detail.

Suppose, then, that the s.d.f. can be represented in the form

$$f(\omega) = B(\omega)B(-\omega) \tag{1}$$

where

$$B(\omega) = \int_0^\infty b_s e^{-is\omega}\, ds \tag{2}$$

is analytic and zero-free in the lower half-plane. If the l.l.s. predictor of $x_{t+\nu}$ is

$$\hat{x}_{t+\nu} = \int_0^\infty \gamma_s x_{t-s}\, ds \tag{3}$$

then

$$\gamma(\omega) = \int_0^\infty \gamma_s e^{-is\omega}\, d\omega = \frac{\int_0^\infty b_{s+\nu} e^{-is\omega}}{B(\omega)} = \frac{[e^{i\nu\omega}B(\omega)]_+}{B(\omega)} \tag{4}$$

and the prediction m.s.e. is

$$\sigma_\nu^2 = \int_0^{\nu-} b_s^2\, ds \tag{5}$$

in analogy with (1.7), (1.8). The weight function γ_s in (3) must be interpreted fairly liberally, since the linear function (3) may involve derivatives of x_t. This will be revealed by polynomial behaviour of $\gamma(\omega)$ (cf. discussion of this point in sections (7.5), (7.6)).

As an example consider the particular case

$$B(\omega) = \frac{\sigma}{L(i\omega)} = \frac{\sigma}{\prod_1^p (i\omega + \alpha_j)} \qquad (6)$$

$$(Re(\alpha_j) > 0, j = 1, 2, \dots p)$$

so that $L(\zeta)$ is a polynomial with all its zeros in the left-half plane. Such a process would be generated by the stochastic differential equation

$$L(D)\int_\tau^t x_s\, ds = \eta_t - \eta_\tau \qquad (7)$$

where η_t is process of uncorrelated increments, an increment over unit time having variance σ^2.

If the roots α_j are distinct then

$$b_s = \sum_j e^{-\alpha_j s} \Big/ \prod_{k \neq j} (\alpha_k - \alpha_j) \qquad (s \geqslant 0) \qquad (8)$$

Substituting into (4) we obtain then

$$\begin{aligned} \gamma(\omega) &= \sum_j e^{-\alpha_j \nu} \prod_{k \neq j} \left(\frac{\alpha_k + i\omega}{\alpha_k - \alpha_j} \right) \\ &= \sum_j \frac{e^{-\alpha_j \nu} L(i\omega)}{(\alpha_j + i\omega) L'(-\alpha_j)} \end{aligned} \qquad (9)$$

a polynomial of degree $p - 1$, indicating that expression (3), which can be written

$$\hat{x}_{t+\nu} = \gamma(-iD) x_t \qquad (10)$$

is a linear combination of the differentials $x_t^{(j)}$, $(j = 0, 1, \dots p - 1)$. In fact, just as before, the prediction is the value that would be obtained by solving

$$L\left(\frac{d}{d\tau}\right) x_\tau = 0 \qquad (\tau \geqslant t) \qquad (11)$$

with initial conditions given by the specified values of $x_t^{(j)}$.

Ex. 1. Show that for $L(\zeta) = \zeta + \alpha$

$$\hat{x}_{t+\nu} = e^{-\alpha\nu} \hat{x}_t$$

$$\sigma_\nu^2 = \frac{\sigma^2}{2\alpha}(1 - e^{-2\alpha\nu})$$

Ex. 2. Show that if

$$L(\zeta) = (\zeta + \lambda + i\mu)(\zeta + \lambda - i\mu)$$

where $\lambda > 0$, then (cf. Ex. 3.3)

$$\hat{x}_{t+\nu} = \mu^{-1} e^{-\lambda\nu} [\mu \cos(\mu\nu) + \lambda \sin(\mu\nu)] x_t + \mu^{-1} e^{-\lambda\nu} \sin(\mu\nu) x'_t$$

Ex. 3. Show that if $L(\zeta) = (\zeta + \alpha)^p$ then

$$\gamma(\omega) = e^{-\nu\alpha} \sum_{j=0}^{p-1} \frac{[\nu(i\omega + \alpha)]^j}{j!},$$

a type of polynomial approximation to $e^{i\nu\omega}$.

Ex. 4. Suppose that $b_s = \begin{bmatrix} 1 & (0 \leqslant s \leqslant \tau) \\ 0 & (s > \tau) \end{bmatrix}$

$f(\omega)$ now has real zeros, which we had assumed not to be the case when deriving (4). Show, nevertheless that if expression (2.1) is written more exactly as $\hat{x}_{t+\nu,\,\nu}$, then formal application of (4) gives

$$\begin{aligned} \hat{x}_{t+\nu,\,\nu} - \hat{x}_{t+\nu-\tau,\,\nu} &= x_t - x_{t+\nu-\tau}, \quad (\nu < \tau) \\ \hat{x}_{t+\nu,\,\nu} &= 0 \quad\quad\quad\quad\quad\quad (\nu \geqslant \tau) \end{aligned}$$

3.6 Continuous Time Processes with Rational Spectral Density Function

Suppose that factorisation (5.1) holds with

$$B(\omega) = \sigma \frac{M(i\omega)}{L(i\omega)} = \sigma \frac{\prod_1^q (i\omega + \beta_j)}{\prod_1^p (i\omega + \alpha_j)} \tag{1}$$

where all α_j, β_j have positive real part. Thus L and M are polynomials of order p and q respectively, and if the process is to have finite variance we must demand that $p > q$.

As emphasised before, the case of a rational s.d.f. is important in view of the fact that an electrical or mechanical filter with a finite number of conventional components will have a rational frequency gain function.

If the α_j are distinct, then we find from (5.4) that

$$\begin{aligned} \gamma(\omega) &= \frac{1}{M(i\omega)} \sum_j e^{-\alpha_j \nu} M(-\alpha_j) \prod_{j \neq k} \left(\frac{\alpha_k + i\omega}{\alpha_k - \alpha_j} \right) \\ &= \frac{L(i\omega)}{M(i\omega)} \sum_j \frac{e^{-\alpha_j \nu} M(-\alpha_j)}{(\alpha_j + i\omega) L'(-\alpha_j)} \end{aligned} \tag{2}$$

Thus, for

$$B(\omega) = \frac{\sigma(\beta + i\omega)}{(\alpha_1 + i\omega)(\alpha_2 + i\omega)} \tag{3}$$

(cf. Ex. 3.8) we have

$$\begin{aligned} \gamma(\omega) &= \frac{e^{-\alpha_1 \nu}(\beta - \alpha_1)(\alpha_2 + i\omega) - e^{-\alpha_2 \nu}(\beta - \alpha_2)(\alpha_1 + i\omega)}{(\beta + i\omega)(\alpha_2 - \alpha_1)} \\ &= c + \frac{d}{\beta + i\omega} \end{aligned} \tag{4}$$

say, so that

$$\hat{x}_{t+\nu} = c x_t + d \int_0^\infty e^{-\beta s} x_{t-s}\, ds \tag{5}$$

Actually, for continuous time it is not so important to obtain the actual predictive relation (5.3) as it is in discrete time, because the prediction will be done by analogue rather than digitally. The transform $\gamma(\omega)$ is itself the important function, as being the gain function characterising the optimum prediction filter.

If $p \leqslant q$ in (1), then the process has infinite variance: it contains "noise" or "noise-derivative" components. This can only be true in some limiting sense: suppose, for example, that $p > q$ in (1), but the coefficients of the higher powers of ω in $L(i\omega)$ are very small. Then, although $f(\omega)$ ultimately converges sufficiently fast for large ω, over an important portion of the frequency spectrum it behaves as if p were less than q, i.e. as if the process contained components with an infinite amount of variation in the higher frequencies.

In this limiting sense, formula (5.4) will give sensible results even if $p \leqslant q$, but the attempt to find a predictor is rather pointless, since the m.s.e. will necessarily be infinite. However, we shall encounter cases in Chapter 6 (the prediction of a signal on the basis of a noise-corrupted signal) where one has to work with a process of infinite variance, and yet can achieve a finite m.s.e.

Ex. 1. Show that if

$$B(\omega) = \frac{i\omega + \beta}{i\omega + \alpha}$$

then

$$\hat{x}_{t+\nu} = (\beta - \alpha)e^{-\alpha\nu}\int_0^\infty e^{-\beta s}x_{t-s}\,ds, \quad (\nu > 0).$$

Ex. 2. Show that if

$$B(\omega) = \frac{(i\omega + \beta)K}{(i\omega + \alpha)(i\omega + K)}$$

then in the limit of infinitely large K the predictor tends to that given in Ex. 1. What happens to the prediction variance?

Ex. 3. Suppose that

$$x_t = x_{1t} + x_{2t} + \ldots + x_{mt}$$

where the component processes are mutually uncorrelated, and x_{jt} follows a scheme of the type (5.7), the operator being of order p_j ($j = 1, 2 \ldots m$). Show that the predictor $\hat{x}_{t+\nu}$ involves derivatives of x_t up to order $\min(p_j) - 1$.

3.7 The Prediction of One Series from Another

Let $\{x_t, y_t\}$ be a bivariate stationary process in discrete time. Consider the calculation of the l.l.s. estimator of y_t from $(x_s; s \leqslant t)$:

$$\hat{y}_t = \sum_0^\infty \gamma_j x_{t-j} \tag{1}$$

This problem includes that of pure prediction, as the special case $y_t = x_{t+\nu}$. It includes many other cases of interest, such as the extraction or prediction of a signal from noise-corrupted observations on the signal, or the estimation of some other linear function, such as a derivative. These applications will be considered in Chapter 6; for the moment we shall content ourselves with proving the generalisation of (2.7):

$$\gamma(z) = \frac{1}{B(z)}\left[\frac{g_{yx}(z)}{B(z^{-1})}\right]_+ \tag{2}$$

where $B(z)$ describes the canonical m.a. form of the x process:

$$g_{xx}(z) = \sigma^2 B(z)B(z^{-1}) \tag{3}$$

$$x_t = \sum_0^\infty b_j \varepsilon_{t-j} \tag{4}$$

Suppose that in terms of the ε's the predictor is

$$\hat{y}_t = \sum_0^\infty \phi_j \varepsilon_{t-j} \tag{5}$$

If

$$c_j = \text{cov}\,(y_t, \varepsilon_{t-j}) \tag{6}$$

then

$$\begin{aligned} E(\hat{y}_t - y_t)^2 &= \text{var}\,(y) - 2\sum \phi_j c_j + \sum \phi_j^2 \\ &= \text{var}\,(y) - \sum c_j^2 + \sum(\phi_j - c_j)^2 \\ &\geqslant \text{var}\,(y) - \sum c_j^2 \end{aligned} \tag{7}$$

equality holding if and only if

$$\phi_j = c_j \qquad (j = 0, 1, 2, \ldots) \tag{8}$$

The l.l.s.e. is thus determined by

$$\begin{aligned} \phi(z) = \sum \phi_j z^j &= [g_{y\varepsilon}(z)]_+ \\ &= \left[\frac{g_{yx}(z)}{B(z^{-1})}\right]_+ \end{aligned} \tag{9}$$

(see (2.9.7)), whence (2) follows. The prediction m.s.e. is given by the last member of (7).

In the manipulations of (7) we assumed that the sums $\sum \phi_j^2$, $\sum c_j^2$ both converged. The justifications are immediate: we have

$$\text{var}\, y - \sum_0^n c_j^2 \geqslant \text{var}\,[y_t - \sum_0^n \phi_j \varepsilon_{t-j}] \geqslant 0 \tag{10}$$

and

$$\begin{aligned} \sigma^2 \sum \phi_j^2 = \text{var}\,\hat{y} &= \text{var}\,[y + (\hat{y} - y)] \\ &\leqslant 2[\text{var}\, y + \text{var}\,(\hat{y} - y)] < \infty \end{aligned} \tag{11}$$

Note that the variate being predicted, y_t, need not be a member of a sequence: it can be a single variate y if we make the convention

$$g_{yx}(z) = \sum_{-\infty}^{\infty} \text{cov}\,(y, x_{t-j}) z^j \tag{12}$$

The derivation given is on the lines of Kolmogorov's treatment; it is in some ways less general than the Wiener method to be given in Chapter 6, just because it is so completely the natural method for this particular problem. The essence is: one "orthogonalises" consecutive x_t, to obtain a new set of variates, the "innovations" ε_t. The variate y is then "projected" on the orthogonal set ε_t, this projection is finally expressed in terms of the original x_t.

The idea of orthogonalising consecutive x_t was first suggested by Wold (1938) who achieved this by fitting autoregressions of indefinitely increasing order. In this way he obtained the representation (2.6.3). Kolmogorov demonstrated the central role of this relation in his definitive treatment of the stationary process [1941a]. His work in prediction [1941b] was based on these ideas. The treatment was much more sophisticated than the one indicated here; he treats deterministic and singular processes, for example. However, the description we have given covers the essential technique.

3.8 Processes with Real Zeros in the Spectral Density Function

We have assumed that $g(z)$ is analytic and zero-free in an annulus $\rho < |z| < \rho^{-1}$. For a rational s.d.f. the only special restriction implied by this is that $g(z)$ should have no zeros on the unit circle, so that $f(\omega)$ should have no real zeros. The point of the condition is to ensure that the process has an autoregressive representation, so that not only may x_t be expressed in terms of $\varepsilon_t, \varepsilon_{t-1} \ldots$, but also that ε_t may be expressed in terms of $x_t, x_{t-1} \ldots$

In fact, the condition will almost always be fulfilled in practice, because the fact that a variable can only be observed with a limited accuracy means that there is a superimposed error, which effectively sets a positive lower bound to $f(\omega)$. Let us nevertheless consider the simplest process for which the condition is violated:

$$x_t = \varepsilon_t - \varepsilon_{t-1} \tag{1}$$

For this process

$$g(z) = \sigma^2(1 - z)(1 - z^{-1}) \tag{2}$$

The obvious autoregressive inversion of (1) would be

$$\sum_{j=0}^{\infty} x_{t-j} = \varepsilon_t \tag{3}$$

and yet the sum in (3) certainly cannot be regarded as the limit of a finite sum $s_n = \sum_0^n x_{t-j}$, for it is readily verified that the sequence s_n does not converge in mean square.

Let us tackle the prediction problem directly by finding the *finite* sum

$$\hat{x}_{t+1}^{(n)} = \sum_0^{n-1} \gamma_j x_{t-j} \tag{4}$$

which approximates x_{t+1} best in mean square. We leave it as an exercise to the reader to show that the optimum predictor is

$$\hat{x}_{t+1}^{(n)} = -\sum_0^{n-1} \left(\frac{n-j}{n+1}\right) x_{t-j} \tag{5}$$

with prediction variance $\frac{n+2}{n+1}\sigma^2$, furthermore that

$$E\left|\sum_0^n\left(\frac{n-j+1}{n+1}\right)x_{t-j} - \varepsilon_t\right|^2 = \frac{2\sigma^2}{n+1} \tag{6}$$

That is, as n tends to infinity the predictor (5) tends to that which would be given by the autoregressive relation (3), but with the left-hand member of (3) interpreted as a *first Césaro sum.* The prediction variance tends to σ^2, as one might hope. Furthermore, (6) shows that the Césaro sum actually tends to ε_t in mean square, i.e. the a.r. relation (3) is valid if only the limit on the left-hand side is taken as a Césaro rather than a direct limit of s_n.

We leave it to the reader to show that ε_t can also be regarded as the limit in mean square of the *Abel sum* $\lim_{\beta\uparrow 1}\sum_0^\infty \beta^j x_{t-j}$. This is statistically meaningful; since if random error were superimposed upon x_t, the a.r. representation of the process would have the form

$$\sum_0^\infty \beta^j x_{t-j} = \varepsilon_t \tag{7}$$

β tending to 1 from below as the variance of the superimposed error decreased to zero.

In fact, neither the Césaro nor the Abel limit can be evaluated in practice: rounding errors on individual x_t would give the limit infinite variance. This is the original contention: that in practice processes with real spectral zeros are not to be found.

Zeros of infinite order can make the process deterministic. This statement is made exact by conditions (2.7.9) and (2.7.10).

3.9 General Comments

All our examples have concerned rational s.d.f.'s. There seems to have been very little work done on anything more general, although almost invariably processes which involve an infinite number of variates (such as a random field in space) will have an s.d.f. which is, indeed, transcendental. To synthesise a filter with irrational transfer function is not easy, and for this reason there is a tendency to use rational approximations.

As an example which may be of some interest, consider prediction of the continuous time process with s.d.f.

$$f(\omega) = \left(1 + \frac{\omega^2}{n}\right)^{-n} \tag{1}$$

which, in the limit of large n, approaches $e^{-\omega^2}$. By criterion (2.7.9) the limiting process should be purely deterministic. In fact,

$$\sigma_\nu^2 = \int_0^\nu b_s^2\, ds = \frac{n^n}{[(n-1)!]^2}\int_0^\nu s^{2n-2}e^{-2s\sqrt{n}}\, ds \tag{2}$$

If we choose n so large that $\nu < \frac{n-1}{\sqrt{n}}$, and the integrand is an increasing function in $(0, \nu)$, then

$$\sigma_\nu^2 < \frac{n^n \nu}{[(n-1)!]^2} \nu^{2n-2} e^{-2\nu\sqrt{n}}$$
$$< \left(\frac{\sigma^2 e^2 \nu}{2\pi}\right)\left(\frac{\nu^2 e^2}{n}\right)^{n-1} e^{-2\nu\sqrt{n}}$$

This tends to zero very fast as n increases, so the process does indeed approach determinism, in the linear sense.

CHAPTER 4

LEAST-SQUARE APPROXIMATION

This chapter is rather loosely related to the rest of the book: it can be read independently of preceding chapters, and the only unavoidable references to it are those in connection with regression sequences (Ch. 8, section 3) and certainty equivalence (Ch. 10, section 7). Nevertheless, it is the only part of the book in which an attempt is made to treat the least-square technique in any generality: for this reason it is important.

4.1 Derivation of the L.L.S. Estimate

We shall assume for the moment that all random variables have zero mean; it is understood throughout that they have finite variance. The cross-covariance matrix of two random vectors **X**, **Y** will be written

$$E(\mathbf{X}\mathbf{Y}^{\dagger}) = \mathbf{V}_{\mathbf{XY}} \tag{1}$$

We write $\mathbf{Y}^{\dagger}$ rather than $\mathbf{Y}'$, since it will occasionally be necessary to work with complex-valued variates. If $\boldsymbol{\xi}$ is an arbitrary vector of constants, then

$$\boldsymbol{\xi}^{\dagger}\mathbf{V}_{\mathbf{XX}}\boldsymbol{\xi} = E|\boldsymbol{\xi}^{\dagger}\mathbf{X}|^2 \geqslant 0 \tag{2}$$

so that $\mathbf{V}_{\mathbf{XX}}$ is positive semi-definite, and can be singular if, and only if, a linear relation holds identically among the elements of **X**.

Consider now the problem of forming the l.l.s. estimate of a variate y from a finite set of variates $x_1, x_2, \ldots x_n$; i.e. of finding

$$\hat{y} = \sum_1^n a_j x_j \tag{3}$$

where the coefficients a_j are to be chosen so that $E(\hat{y} - y)^2$ is a minimum. In preparation for the multivariate case, to come later, we shall write (3) in the matrix form

$$\hat{\mathbf{Y}} = \mathbf{A}\mathbf{X} \tag{4}$$

despite the fact that $\hat{\mathbf{Y}}$ (or **Y**) is only a 1×1 matrix with the single element $\hat{y}$ (or y).

If $\boldsymbol{\delta} = \hat{\mathbf{Y}} - \mathbf{Y}$, then

$$\begin{aligned} E(\boldsymbol{\delta}\boldsymbol{\delta}^{\dagger}) &= E(\mathbf{AX} - \mathbf{Y})(\mathbf{AX} - \mathbf{Y})^{\dagger} \\ &= \mathbf{A}\mathbf{V}_{\mathbf{XX}}\mathbf{A}^{\dagger} - \mathbf{A}\mathbf{V}_{\mathbf{XY}} - \mathbf{V}_{\mathbf{YX}}\mathbf{A}^{\dagger} + \mathbf{V}_{\mathbf{YY}} \\ &= (\mathbf{A} - \mathbf{V}_{\mathbf{YX}}\mathbf{V}_{\mathbf{XX}}^{-1})\mathbf{V}_{\mathbf{XX}}(\mathbf{A} - \mathbf{V}_{\mathbf{YX}}\mathbf{V}_{\mathbf{XX}}^{-1})^{\dagger} + \mathbf{V}_{\mathbf{YY}} - \mathbf{V}_{\mathbf{YX}}\mathbf{V}_{\mathbf{XX}}^{-1}\mathbf{V}_{\mathbf{XY}} \end{aligned} \tag{5}$$

In forming the final member of (5) we have assumed $\mathbf{V}_{\mathbf{XX}}$ non-singular. This is hardly any restriction in the case of finite n, since redundant variates can always be dropped from the set $x_1 \ldots x_n$ until one is left with a set among which no linear relations exist. The situation may be less simple if there are infinitely many variates.

Since $\mathbf{V_{XX}}$ is positive definite, expression (5) will attain its unique minimum when

$$\mathbf{A} = \mathbf{V_{YX}V_{XX}^{-1}} \tag{6}$$

so that the l.l.s. estimate of $\mathbf{Y}$ is

$$\hat{\mathbf{Y}} = \mathbf{V_{YX}V_{XX}^{-1}X} \tag{7}$$

and the prediction m.s.e.

$$\mathbf{V_{\delta\delta}} = E(\boldsymbol{\delta}\boldsymbol{\delta}^{\dagger}) = \mathbf{V_{YY}} - \mathbf{V_{YX}V_{XX}^{-1}V_{XY}} \tag{8}$$

In order to make (7) explicit in any particular case, only one problem remains: to invert the covariance matrix $\mathbf{V_{XX}}$ (or some corresponding linear operator, in more general cases). The extensive literature on prediction theory boils down to little more than this: the inversion of a linear operator.

We shall sometimes refer to $\hat{\mathbf{Y}}$ as the projection of $\mathbf{Y}$ on $\mathbf{X}$ (more properly, of $\mathbf{Y}$ on the space of $\mathbf{X}$); this geometric analogy is often graphic and convenient.

Ex. 1. The equation $\mathbf{AV_{XX}} = \mathbf{V_{YX}}$ can be rewritten

$$\operatorname{cov}(\hat{y} - y, x_j) = 0 \quad (\text{all } x_j)$$

Show that this condition is necessary for a minimum m.s.e., even if j ranges over an infinite set, or if linear relations exist among the x_j.

Ex. 2. Show that (8) can be written in the alternative forms

$$\mathbf{V_{\delta\delta}} = \mathbf{V_{YY}} - \mathbf{AV_{XY}} = \mathbf{V_{YY}} - \mathbf{AV_{XX}A^{\dagger}}$$
$$= \begin{vmatrix} \mathbf{V_{YY}} & \mathbf{V_{YX}} \\ \mathbf{V_{XY}} & \mathbf{V_{XX}} \end{vmatrix} \Big/ |\mathbf{V_{XX}}|$$

and also as

$$\mathbf{V_{YY}} = \mathbf{V_{\hat{Y}\hat{Y}}} + \mathbf{V_{\delta\delta}}$$

Suppose that one is trying to approximate several variables simultaneously, so that $\mathbf{Y}$ is a column-vector with elements $y_1, y_2, \ldots y_m$. Then we see from (5) that (7) is still the solution, in the sense that this choice minimises the m.s.e. of any linear function of the y's, $\boldsymbol{\xi}^{\dagger}\mathbf{Y}$. In particular, the estimates of the individual y's are just those given by the single-variable solution. Expression (8) is now an $m \times m$ matrix, the covariance matrix of estimation errors.

Ex. 3. Show that $E(\boldsymbol{\delta}\mathbf{X}^{\dagger}) = 0$; i.e. the elements of $\mathbf{X}$ are uncorrelated with those of $\boldsymbol{\delta}$.

Ex. 4. Show that the choice (7) also minimises $E(\boldsymbol{\delta}^{\dagger}\mathbf{K}\boldsymbol{\delta})$ and $|E(\boldsymbol{\delta}\boldsymbol{\delta}^{\dagger})|$, where $\mathbf{K}$ is any positive definite matrix.

Ex. 5. Show that the l.l.s.e. of $\sum_1^m \xi_j y_j$ is $\sum_1^m \xi_j \hat{y}_j$, the $\hat{y}_j$ being the individual l.l.s.e. of the y_j.

Ex. 6. Show that

$$\begin{pmatrix} \mathbf{X} \\ \mathbf{Y} \end{pmatrix}^{\dagger} \begin{pmatrix} \mathbf{V_{XX}} & \mathbf{V_{XY}} \\ \mathbf{V_{YX}} & \mathbf{V_{YY}} \end{pmatrix}^{-1} \begin{pmatrix} \mathbf{X} \\ \mathbf{Y} \end{pmatrix} = \mathbf{X}^{\dagger}\mathbf{V_{XX}^{-1}X} + (\mathbf{Y} - \hat{\mathbf{Y}})^{\dagger}\,\mathbf{V_{\delta\delta}^{-1}}\,(\mathbf{Y} - \hat{\mathbf{Y}})$$

$\hat{\mathbf{Y}}$ and $\mathbf{V}_{\delta\delta}$ being defined by (7), (8). This implies that $\hat{\mathbf{Y}}$ is identical with the expression for the conditional expectation of $\mathbf{Y}$ for a given $\mathbf{X}$, calculated on the assumption that $\mathbf{X}$ and $\mathbf{Y}$ are jointly normally distributed.

4.2 Methods of Inverting a Covariance Matrix

The simplest case of all, in which no problem of matrix inversion arises, is that in which all the x_j are uncorrelated and have unit variance, so that $\mathbf{V_{xx}} = \mathbf{I}$. In this case we see from (1.7) and (1.8) (or directly) that, for estimation of a single variable, y,

$$\hat{y} = \sum_1^n \gamma_j x_j \tag{1}$$

$$E(\delta^2) = \text{var}(y) - \sum_1^n \gamma_j^2 \tag{2}$$

where

$$\gamma_j = \text{cov}(y, x_j) \tag{3}$$

If further uncorrelated variates, $x_{n+1}, x_{n+2}, \ldots$, are brought into the estimate, it is plain from (1) that the coefficients of the old x_j remain unchanged. Furthermore, since expression (2) is non-negative, then if n is allowed to tend to infinity we shall still have

$$\sum_1^\infty \gamma_j^2 < \infty \tag{4}$$

This situation is too good to be common. One treats the general case, in so far as it can be treated explicitly, by representing the x_j as linear functions of an uncorrelated sequence. Two such *orthogonal representations* are associated with two particular representations of the covariance matrix: the triangular and the spectral representations. In the case of stationary process prediction these two representations correspond to study of the process in the time or in the frequency domain, and largely correspond to the methods of Kolmogorov and of Wiener, respectively.

Suppose one carries out a *Gram-Schmidt orthonormalisation* of the variates x_j by constructing the linear forms

$$\varepsilon_j = \sum_{k=1}^{j} c_{jk} x_k \qquad (j = 1, 2, \ldots) \tag{5}$$

the c_{jk} being chosen so that

$$E(\varepsilon_j \varepsilon_k) = \delta_{jk} \tag{6}$$

This process can be continued as long as no linear relation exists among the x_j; it implies that

$$E(\varepsilon_j x_k) = 0 \qquad (j > k) \tag{7}$$

The system (5) can be inverted recursively to a set of relations of the form

$$x_j = \sum_{k=1}^{j} d_{jk} \varepsilon_k \tag{8}$$

Equation (8) is an orthogonal representation of the variate ((5) and (8) correspond in fact to the a.r. and m.a. representations of a stationary process); the corresponding representation of the covariance is evidently

$$\mathbf{V_{xx}} = \mathbf{DD}^{\dagger} \tag{9}$$

where $\mathbf{D}$ is the lower triangular matrix with elements d_{jk}.

If y may depend upon $x_1, x_2, \ldots x_n$, then we see from (1) and (5) that

$$\begin{aligned}\hat{y} &= \sum_{j=1}^{n} \operatorname{cov}(y, \varepsilon_j)\varepsilon_j \\ &= \sum_{j=1}^{n} \sum_{k=1}^{j} \sum_{l=1}^{j} c_{jk}c_{jl} \operatorname{cov}(y, x_k)x_l \end{aligned} \tag{10}$$

For finite n this is a perfectly practicable way of calculating $\hat{y}$: effectively, we have inverted $\mathbf{V_{xx}}$ by the method of *triangular resolution* (see Hartree (1952), p. 164). The method is a particularly useful one if there is a natural ordering of the variates x_j and if one wishes to increase n successively.

The other important method is based on the *spectral representation* of the Hermitian matrix $\mathbf{V_{xx}}$:

$$\mathbf{V_{xx}} = \sum_{\nu=1}^{n} \lambda_\nu \boldsymbol{\xi}_\nu \boldsymbol{\xi}_\nu^{\dagger} \tag{11}$$

Here the λ_ν and $\boldsymbol{\xi}_\nu$ are the eigenvalues and eigenvectors of the matrix, characterised by

$$\mathbf{V_{xx}}\boldsymbol{\xi}_\nu = \lambda_\nu \boldsymbol{\xi}_\nu \tag{12}$$

One can always standardise the $\boldsymbol{\xi}_\nu$ so that

$$\boldsymbol{\xi}_\mu^{\dagger}\boldsymbol{\xi}_\nu = \delta_{\mu\nu} \tag{13}$$

and the $n \times n$ matrix whose column vectors are the $\boldsymbol{\xi}_\nu$ is unitary.

Defining the scalar variate

$$\zeta_\nu = \boldsymbol{\xi}_\nu^{\dagger}\mathbf{X} \tag{14}$$

one has

$$\mathbf{X} = \sum_{\nu=1}^{n} \zeta_\nu \boldsymbol{\xi}_\nu \tag{15}$$

$$E(\zeta_\mu \bar{\zeta}_\nu) = \boldsymbol{\xi}_\mu^{\dagger}\mathbf{V_{xx}}\boldsymbol{\xi}_\nu = \lambda_\nu \delta_{\mu\nu} \tag{16}$$

so that (15) is another orthogonal representation of the variate, the *spectral representation*.

This particular representation has several features which distinguish it; for example, the variates ζ_ν are the principal components of $\mathbf{X}$ (see for example, Wilks (1962), p. 564), the linear forms $\boldsymbol{\alpha}^{\dagger}\mathbf{X}$ having extremal variance for a given $\boldsymbol{\alpha}^{\dagger}\boldsymbol{\alpha}$. However, the most important point is that, in many cases, the symmetry properties of the process $\mathbf{X}$ alone will determine the eigenvectors $\boldsymbol{\xi}_\nu$. For example, the fact that a stationary process is invariant under time translation determines the spectral representation as being a Fourier representation (see the example of the circulant process, below).

From either the orthogonal representation (15) or the general formula (1.7) we find that

$$\hat{\mathbf{Y}} = \sum \lambda_\nu^{-1} \mathbf{V}_{\mathbf{YX}} \xi_\nu \xi_\nu^\dagger \mathbf{X} \tag{17}$$

Ex. 1. Show that if $\lambda_\nu = 0$ then arbitrary weight may be given to the νth term in sum (17), for it is zero.

As an example, consider the important case of the real *circulant process,* characterised by the fact that a cyclic permutation of $(x_1, x_2, \ldots x_n)$ leaves the covariance matrix unchanged. Thus, for any integral τ

$$\text{cov}\,(x_s, x_t) = \text{cov}\,(x_{s+\tau}, x_{t+\tau}) \tag{18}$$

where the subscripts of the variates are always reduced modulo n to one of the integers $1, 2, \ldots n$. We thus find that we can write

$$\text{cov}\,(x_s, x_t) = \Gamma_{s-t} \tag{19}$$

where the function Γ_s is even, and has period n. This process is the finite equivalent of a stationary process, invariance under translation along an infinite axis being replaced by invariance under translation along a periodic or "loop" axis.

One could characterise the situation by saying that **TX** has the same covariance matrix as **X**, **T** being the matrix

$$\mathbf{T} = \begin{bmatrix} \cdot & 1 & \cdot & \ldots & \cdot \\ \cdot & \cdot & 1 & \ldots & \cdot \\ & & \ldots & & \\ \cdot & \cdot & \cdot & \ldots & 1 \\ 1 & \cdot & \cdot & \ldots & \cdot \end{bmatrix} \tag{20}$$

which effects a simple cyclic permutation.

Thus

$$\mathbf{V}_{\mathbf{XX}} = \mathbf{T}\mathbf{V}_{\mathbf{XX}}\mathbf{T}' \tag{21}$$

or

$$\mathbf{V}_{\mathbf{XX}}\mathbf{T} = \mathbf{T}\mathbf{V}_{\mathbf{XX}} \tag{22}$$

since **T** is a permutation matrix, and so orthogonal. From the fact that $\mathbf{V}_{\mathbf{XX}}$ and **T** commute we deduce that they have the same set of eigenvectors, and the right eigenvectors of **T** are readily found to be given by

$$\xi_\nu = \frac{1}{\sqrt{n}} \begin{bmatrix} \theta^\nu \\ \theta^{2\nu} \\ \cdot \\ \cdot \\ \cdot \\ \theta^{n\nu} \end{bmatrix} \tag{23}$$

where

$$\theta = e^{2\pi i/n} \tag{24}$$

These are also the eigenvectors of $\mathbf{V_{xx}}$, with corresponding eigenvalues

$$\lambda_\nu = \sum_{s=1}^{n} \Gamma_s \theta^{-\nu s} \tag{25}$$

as can be directly verified. The spectral representations of the covariance and the variate, (11) and (12), are, in this case

$$\Gamma_s = \frac{1}{n} \sum_{\nu=1}^{n} \lambda_\nu \theta^{\nu s} \tag{26}$$

$$x_s = \frac{1}{\sqrt{n}} \sum_{\nu=1}^{n} \zeta_\nu \theta^{\nu s} \tag{27}$$

where

$$\zeta_\nu = \boldsymbol{\xi}_\nu^\dagger \mathbf{X} = \frac{1}{\sqrt{n}} \sum_{s=1}^{n} x_s \theta^{-\nu s} \tag{28}$$

That is, the spectral representations are finite Fourier representations, and the eigenvalues λ_ν constitute the "spectral density function" of this "finite stationary process". For the coefficients a_j of (1.3) one can say that

$$(\sum a_j \theta^{\nu j}) = \frac{1}{\lambda_\nu}(\sum \operatorname{cov}(y, x_j) \theta^{\nu j}) \tag{29}$$

The point of this example is that statistical invariance of $\mathbf{X}$ under the operation $\mathbf{T}$ completely determines the eigenvectors $\boldsymbol{\xi}_\nu$, and results in the spectral representations being determined as finite Fourier representations. The analogy with the conventional stationary process is plain.

One can see in this special case the essentials of a more general situation: there is a set of permutation operators on $\mathbf{X}$ (in this case the cyclic permutations) which leave $\mathbf{V}$ unchanged, and which constitute the elements of a group. Furthermore, the natural transformations of $\mathbf{X}$ to consider are just linear combinations of these permutations (corresponding to m.a. transformations of a process in the stationary case).

4.3 The Inclusion of Deterministic Terms in the Least-square Approximation

Suppose now that the vector variates $\mathbf{X}$ and $\mathbf{Y}$ do not have zero mean, but that

$$E(\mathbf{X}) = \sum_{1}^{q} \beta_j \mathbf{G}_j = \mathbf{G}\boldsymbol{\beta} \tag{1}$$

$$E(\mathbf{Y}) = \sum_{1}^{q} \beta_j \mathbf{H}_j = \mathbf{H}\boldsymbol{\beta} \tag{2}$$

where the β_j are unknown scalars ("regression coefficients"), forming a $q \times 1$ vector $\boldsymbol{\beta}$, while $\mathbf{G}$ and $\mathbf{H}$ are matrices whose columns are known sequences, $\mathbf{G}_j$ and $\mathbf{H}_j$.

As an example, suppose that one is predicting $x_{n+\nu}$ on the basis of a sample $x_1, x_2, \ldots x_n$ from a process $\{x_t\}$ having the representation

$$x_t = \sum_{j=1}^{q} \beta_j G_j(t) + u_t \tag{3}$$

where the $G_j(t)$ are known functions of time, and u_t a purely non-deterministic stationary process with known auto-covariances. Then $\mathbf{G}_j$ has the elements $G_j(t)$ $(t = 1, 2, \ldots n)$ and $\mathbf{H}_j$ has the single element $G_j(n + \nu)$. It may seem unreasonable to suppose the β_j unknown, but the structure of $\{u_t\}$ known. The reasons for this assumption will be discussed in Ch. 8.

For simplicity, let us temporarily revert to the prediction of a single variate, so that $\mathbf{Y}$ has only one element. We shall now seek the linear predictor (1.4) determined by

$$E(\hat{\mathbf{Y}} - \mathbf{Y})(\hat{\mathbf{Y}} - \mathbf{Y})^{\dagger} = \min_{\mathbf{A}} \max_{\beta} E(\mathbf{AX} - \mathbf{Y})(\mathbf{AX} - \mathbf{Y})^{\dagger} \tag{4}$$

Our former l.s. criterion must be modified in some way to take account of the unknown constants, β_j, and the minimax criterion is a reasonable and convenient choice.

Now, since for any quadratic function $Q(\mathbf{X})$ we have

$$EQ(\mathbf{X}) = EQ[\mathbf{X} - E(\mathbf{X})] + Q[E(\mathbf{X})] \tag{5}$$

then

$$E(\mathbf{AX} - \mathbf{Y})(\mathbf{AX} - \mathbf{Y})^{\dagger} = \mathbf{AV_{XX}A}^{\dagger} - \mathbf{AV_{XY}} - \mathbf{V_{YX}A}^{\dagger} + \mathbf{V_{YY}} + (\mathbf{AG} - \mathbf{H})\boldsymbol{\beta}\boldsymbol{\beta}^{\dagger}(\mathbf{AG} - \mathbf{H})^{\dagger} \tag{6}$$

The last term in (6) is the square of a scalar, $|(\mathbf{AG} - \mathbf{H})\beta|^2$. Evidently the maximum of this with respect to $\boldsymbol{\beta}$ will be indefinitely large unless

$$\mathbf{AH} = \mathbf{G} \tag{7}$$

and this is evidently a condition which must be imposed if the minimum in (4) is even to be finite.

The conclusion is, then, that the quantity to be minimised is just that considered before, (1.5), but that the minimisation is now subject to a side-condition (7), (in fact, q side-conditions). The interpretation of the condition is that the least-square approximation should be *exact* if $\mathbf{X}$ and $\mathbf{Y}$ are of the deterministic form $\sum \beta_j \mathbf{G}_j, \sum \beta_j \mathbf{H}_j$, whatever the β_j. Some readers might find this demand so reasonable that they could have accepted it as a starting-point. However, I find it more instructive to regard condition (7) as a consequence of the minimax criterion (4), rather than as a self-evident requirement.

The quantity to be minimised can now be taken as

$$\mathbf{AV_{XX}A}^{\dagger} - \mathbf{AV_{XY}} - \mathbf{V_{YX}A}^{\dagger} + \mathbf{V_{YY}} + (\mathbf{H} - \mathbf{AG})\mathbf{K} + \mathbf{K}^{\dagger}(\mathbf{H} - \mathbf{AG})^{\dagger} \tag{8}$$

where $\mathbf{K}$ is a vector (or, in the case of multivariate y, a matrix) of Lagrangian multipliers. Just as (1.6) minimised expression (1.5), so expression (8) is minimised with respect to $\mathbf{A}$ by the choice

$$\mathbf{A} = (\mathbf{V_{YX}} + \mathbf{K}^\dagger\mathbf{G}^\dagger)\mathbf{V_{XX}^{-1}} \tag{9}$$

Evaluating $\mathbf{K}$ by substitution of expression (9) for $\mathbf{A}$ into equation (7), and substituting the resulting value of $\mathbf{A}$ into (1.4), we find that the predictor $\hat{\mathbf{Y}}$ satisfying the extended l.l.s principle (4) can be written

$$\hat{\mathbf{Y}} = \mathbf{H}\hat{\boldsymbol{\beta}} + \mathbf{V_{YX}}\mathbf{V_{XX}^{-1}}(\mathbf{X} - \mathbf{G}\hat{\boldsymbol{\beta}}) \tag{10}$$

where

$$\hat{\boldsymbol{\beta}} = (\mathbf{G}^\dagger\mathbf{V_{XX}^{-1}}\mathbf{G})^{-1}\mathbf{G}^\dagger\mathbf{V_{XX}^{-1}}\mathbf{X} \tag{11}$$

That is, the vectors $\mathbf{X}$ and $\mathbf{Y}$ are corrected for their mean values (1) and (2), the unknown coefficient vector $\boldsymbol{\beta}$ being estimated by (11). The prediction formula (1.7) that was formerly used on $\mathbf{X}$ and $\mathbf{Y}$ directly is then applied to the estimated deviations from the mean values.

We find from (10), (11) that the m.s.e. is now

$$\mathbf{V}_{\delta\delta} = (\mathbf{V_{YY}} - \mathbf{V_{YX}}\mathbf{V_{XX}^{-1}}\mathbf{V_{XY}}) + (\mathbf{H} - \mathbf{V_{YX}}\mathbf{V_{XX}^{-1}}\mathbf{G})(\mathbf{G}^\dagger\mathbf{V_{XX}^{-1}}\mathbf{G})^{-1}(\mathbf{H} - \mathbf{V_{YX}}\mathbf{V_{XX}^{-1}}\mathbf{G})^\dagger \tag{12}$$

The extension to the case of multivariate y goes just as before, equations (10) and (12) then being genuine matrix equations, and the extremal characterisation (4) now holding if both sides of the equation are post-multiplied by any vector ξ, and pre-multiplied by the corresponding $\xi^\dagger$.

A statistician will recognise the $\hat{\boldsymbol{\beta}}$ of equation (11) as being just the maximum likelihood estimate of $\boldsymbol{\beta}$, based on the sample vector $\mathbf{X}$, if this vector is assumed normally distributed. In fact, if $\mathbf{X}$ and $\mathbf{Y}$ are jointly normally distributed, then the logarithm of their joint likelihood is proportional to

$$(\mathbf{X} - \mathbf{G}\boldsymbol{\beta})^\dagger\mathbf{V_{XX}^{-1}}(\mathbf{X} - \mathbf{G}\boldsymbol{\beta}) + \mathbf{Z}^\dagger\mathbf{W}^{-1}\mathbf{Z} \tag{13}$$

where

$$\mathbf{Z} = \mathbf{Y} - \mathbf{H}\boldsymbol{\beta} - \mathbf{V_{YX}}\mathbf{V_{XX}^{-1}}(\mathbf{X} - \mathbf{G}\boldsymbol{\beta}) \tag{14}$$

and $\mathbf{W}$ is the matrix (1.8). Minimising this with respect to $\boldsymbol{\beta}$ and $\mathbf{Y}$ we obtain just the expressions (10) and (11).

In order to evaluate the l.l.s. estimate (10) explicitly, two calculations are necessary: the inversion of the matrices $\mathbf{V_{XX}}$ and $\mathbf{G}^\dagger\mathbf{V_{XX}^{-1}}\mathbf{G}$, of orders n and q respectively. Since q will generally be relatively small, the only real problem is, as before, the inversion of the covariance matrix $\mathbf{V_{XX}}$.

4.4 The Minimisation of More General Quadratic Forms

We have considered the problem of finding a linear function of $\mathbf{X}$, $\hat{\mathbf{Y}}$, which minimises the particular quadratic form $E|\hat{\mathbf{Y}} - \mathbf{Y}|^2$. In Ch. 10 we shall find it necessary to work with criterion functions that are rather general quadratic forms in the variates, and we shall now deduce some preparatory results.

Consider two vector variables, $\boldsymbol{\xi}$ and $\mathbf{X}$, where the elements of $\mathbf{X}$ may be random variables. We wish to choose $\boldsymbol{\xi}$ so as to minimise a given quadratic form $Q(\boldsymbol{\xi}, \mathbf{X})$, this form being positive definite in $\boldsymbol{\xi}$. The minimising $\boldsymbol{\xi}$ is obviously a linear function of $\mathbf{X}$: in fact, we can complete the square and write

$$Q(\boldsymbol{\xi}, \mathbf{X}) = (\boldsymbol{\xi} - \mathbf{AX})'\mathbf{P}(\boldsymbol{\xi} - \mathbf{AX}) + \ldots \tag{1}$$

where $\mathbf{P}$ is positive definite and the dots indicate terms not containing $\boldsymbol{\xi}$. The function $\mathbf{Q}$ is then minimised by the choice $\boldsymbol{\xi} = \mathbf{AX}$.

Suppose we consider now the problem of minimising

$$\bar{Q} = E[Q(\boldsymbol{\xi}, \mathbf{X})] \tag{2}$$

with respect to $\boldsymbol{\xi}$; $\boldsymbol{\xi}$ restricted to being a linear function of $\mathbf{X}$. Then, a statement which is almost trivial, and yet is best made explicitly at some stage, is that the solution will be identical with that for the free minimisation of Q itself, namely, $\boldsymbol{\xi} = \mathbf{AX}$. We see from (1) that this choice will minimise expression (2) absolutely, and it is of the required linear form.

Consider now a quadratic form with three vector arguments, $Q(\boldsymbol{\xi}, \mathbf{X}, \mathbf{Y})$, positive definite in $\boldsymbol{\xi}$. The vectors $\mathbf{X}$ and $\mathbf{Y}$ are supposed random with zero means and finite variances, but otherwise of unrestricted distribution. Our main result can then be enunciated as follows:

Theorem 1

The vector $\boldsymbol{\xi}$ *which minimises*

$$\bar{Q} = E[Q(\boldsymbol{\xi}, \mathbf{X}, \mathbf{Y})] \tag{3}$$

subject to the restriction that $\boldsymbol{\xi}$ *be linearly dependent upon* $\mathbf{X}$ *and functionally independent of* $\mathbf{Y}$, *is identical with the* $\boldsymbol{\xi}$ *that freely minimises*

$$Q(\boldsymbol{\xi}, \mathbf{X}, \mathbf{V}_{\mathbf{YX}}\mathbf{V}_{\mathbf{XX}}^{-1}\mathbf{X}) \tag{4}$$

i.e., *the original form with* $\mathbf{Y}$ *replaced by* $\hat{\mathbf{Y}}$, *its l.l.s. approximation in terms of* $\mathbf{X}$.

To prove this result we note that we can write

$$\mathbf{Y} = \mathbf{V}_{\mathbf{YX}}\mathbf{V}_{\mathbf{XX}}^{-1}\mathbf{X} + \boldsymbol{\delta} \tag{5}$$

where the elements of $\boldsymbol{\delta}$ are uncorrelated with those of $\mathbf{X}$ (see Ex. 1.3). Expansion of Q about $\boldsymbol{\delta} = 0$ gives

$$Q(\boldsymbol{\xi}, \mathbf{X}, \mathbf{Y}) = Q(\boldsymbol{\xi}, \mathbf{X}, \mathbf{V}_{\mathbf{YX}}\mathbf{V}_{\mathbf{XX}}^{-1}\mathbf{X}) + \boldsymbol{\delta}'\mathbf{R}\boldsymbol{\xi} + \ldots \tag{6}$$

where $\mathbf{R}$ is a constant matrix, and the dots indicate terms not containing $\boldsymbol{\xi}$. Now, $\boldsymbol{\xi}$ is a linear function of $\mathbf{X}$, and thus also uncorrelated with $\boldsymbol{\delta}$, so that $E(\boldsymbol{\delta}'\mathbf{R}\boldsymbol{\xi}) = 0$. We thus have

$$\bar{Q} = E[Q(\boldsymbol{\xi}, \mathbf{X}, \mathbf{V}_{\mathbf{YX}}\mathbf{V}_{\mathbf{XX}}^{-1}\mathbf{X})] + (\text{terms not containing } \boldsymbol{\xi}) \tag{7}$$

The theorem now follows from the principle stated after equation (2).

If the random arguments of Q have non-zero means, then the result is modified only slightly.

Theorem 2

Suppose that $\bar{\boldsymbol{\xi}}$ *and* $\boldsymbol{\xi}$ *are to be chosen so as to minimise*

$$\bar{Q} = E[Q(\bar{\boldsymbol{\xi}} + \boldsymbol{\xi}, \bar{\mathbf{X}} + \mathbf{X}, \bar{\mathbf{Y}} + \mathbf{Y})] \tag{8}$$

where $\bar{\boldsymbol{\xi}}$, $\bar{\mathbf{X}}$ *and* $\bar{\mathbf{Y}}$ *are constant vectors,* $\mathbf{X}$ *and* $\mathbf{Y}$ *have zero means, and* $\boldsymbol{\xi}$ *is restricted to being a linear function of* $\mathbf{X}$. *Then* $\bar{\boldsymbol{\xi}}$ *is chosen as the free minimiser of* $Q(\bar{\boldsymbol{\xi}}, \bar{\mathbf{X}}, \bar{\mathbf{Y}})$ *and* $\boldsymbol{\xi}$ *as the free minimiser of expression* (4).

This follows immediately from the fact that expression (8) equals

$$\bar{Q} = Q(\bar{\boldsymbol{\xi}}, \bar{\mathbf{X}}, \bar{\mathbf{Y}}) + E[Q(\boldsymbol{\xi}, \mathbf{X}, \mathbf{Y})] \tag{9}$$

since if Q is expanded about $\bar{\boldsymbol{\xi}}$, $\bar{\mathbf{X}}$, $\bar{\mathbf{Y}}$ and an expectation taken, terms linear in the random variables will vanish.

CHAPTER 5

PROJECTION ON THE INFINITE SAMPLE

We shall now consider the l.l.s. estimation of a variate y on the basis of a complete realisation of a stationary process $\{x_t\}$. Thus, the estimate $\hat{y}$ may be based upon the "future" as well as the "past" of x: this corresponds to the construction of what is sometimes known as an *infinite lag filter*. The calculation is a useful preliminary to the one in which y is to be based upon the "past" of x_t alone, touched on already in Ch. 3, and to be considered further in Ch. 6. However, the case is of interest in its own right, for both theory and application.

The only essential prerequisite is Ch. 2, although occasional references will be made to Ch. 4.

5.1 Construction of the Estimate

As examples of the sort of application in mind, one could mention the "smoothing" of the observed x_t series in an attempt to extract some component process, y_t; or the interpolation of the series in order to estimate an unobserved value, $y = x_{t+\nu}$ $(0 < \nu < 1)$.

We may be attempting to estimate a particular term y_t in a process $\{y_t\}$ which is jointly stationary with $\{x_t\}$; alternatively, y may be a single variable, not to be regarded as a member of a sequence. There is little difference between the two cases: we have already extended our definition of $g_{yx}(z)$ or $f_{yx}(\omega)$ in the first case (formula (2.1.19) to the second (formula (3.7.12)).

Let us consider the discrete case first, and suppose that

$$\hat{y} = \sum_{-\infty}^{\infty} \gamma_s x_{t-s} \tag{1}$$

$$\gamma(z) = \sum \gamma_s z^s \tag{2}$$

Then, by formal generalisation of the circulant example of Ch. 4 (see equation (4.2.29)), one might expect that

$$\gamma(z) = \frac{g_{yx}(z)}{g_{xx}(z)} = \frac{f_{yx}(\omega)}{f_{xx}(\omega)} \tag{3}$$

with all expansions valid at least on $|z| = 1$. Formula (3) will in fact be valid if g_{yx} and g_{xx} are convergent, and g_{xx} non-zero, on $|z| = 1$. However, a more general solution can be obtained relatively easily.

We have (see Ex. (4.1.1))

$$\begin{aligned} 0 &= \operatorname{cov}(y - \hat{y}, x_{t-j}) \\ &= \operatorname{cov}(y, x_{t-j}) - \sum \gamma_k \Gamma_{j-k} \\ &= \frac{1}{2\pi}\int_{-\pi}^{\pi} e^{ij\omega}[dF_{yx}(\omega) - \gamma(e^{-i\omega})dF_{xx}(\omega)] \qquad (4) \\ &\qquad\qquad (j = 0, \pm 1, \pm 2, \ldots) \end{aligned}$$

The fact that $\hat{y}$ will have finite variance ensures the convergence of the integral (4).

Now, since (4) holds for all integral j, we must have

$$dF_{yx}(\omega) - \gamma(e^{-i\omega})dF_{xx}(\omega) = 0 \tag{5}$$

almost everywhere. Thus

$$\gamma(e^{-i\omega}) = \frac{dF_{yx}(\omega)}{dF_{xx}(\omega)} \tag{6}$$

for almost all ω for which $dF_{xx}(\omega) > 0$. If $dF_{xx} = 0$ for $\omega = \lambda$ then $\gamma(e^{-i\lambda})$ is undefined by (5), and in fact its value is immaterial, because x_t cannot then contain any Fourier component of frequency λ. We find that

$$E(\delta^2) = E(\hat{y} - y)^2 = \operatorname{var}(y) - \frac{1}{2\pi}\int \frac{|dF_{yx}(\omega)|^2}{dF_{xx}(\omega)} \tag{7}$$

$$dF_{\delta\delta}(\omega) = dF_{yy}(\omega) - \frac{|dF_{yx}(\omega)|^2}{dF_{xx}(\omega)} \tag{7'}$$

the second formula being appropriate if y is an element of a stationary process $\{y_t\}$.

In most cases we shall be dealing with processes with well-behaved spectral densities, for which solution (6) reduces to (3), and (7) has an obvious analogous expression. However, if x_t contains pure harmonic components, for example, then we must revert to the solution (6). It is interesting that the "almost everywhere" proviso associated with this solution may be important. If the right-hand member of (6) cannot be represented everywhere by its Fourier series then this may actually affect the construction of $\hat{y}$ and the value of its m.s.e.: see the discussion after Ex. (2.3).

A particularly important special case is that of *signal extraction*, where

$$x_t = y_t + \eta_t \tag{8}$$

and the interpretation is that y is a signal which one wishes to read, and η a superimposed error, so that the observed value x is the noise-corrupted signal. Assuming the existence of well-behaved spectral densities, we have then from (3)

$$\gamma(z) = \frac{f_{yy} + f_{y\eta}}{f_{yy} + f_{y\eta} + f_{\eta y} + f_{\eta\eta}} \tag{9}$$

The spectral density of the error, $\delta = \hat{y} - y$, is readily found to be given by

$$f_{\delta\delta} = \frac{f_{yy}f_{\eta\eta} - f_{y\eta}f_{\eta y}}{f_{yy} + f_{y\eta} + f_{\eta y} + f_{\eta\eta}} \tag{10}$$

If the signal and error processes are mutually uncorrelated, then (9) and (10) reduce to

$$\gamma(z) = \frac{f_{yy}}{f_{yy} + f_{\eta\eta}} \tag{11}$$

$$f_{\delta\delta} = \frac{f_{yy}f_{\eta\eta}}{f_{yy} + f_{\eta\eta}} \tag{12}$$

We see from (11) that the estimation rule in this case can be characterised very simply: one multiplies a particular frequency component of $\{x_t\}$ by the fraction of the sampling variance of this component that can be attributed to the signal $\{y_t\}$. Formula (12) is also informative: if there is no overlap in the frequency spectra of the signal and error processes, i.e. no ω value for which both f_{yy} and $f_{\eta\eta}$ are non-zero, then $f_{\delta\delta} \equiv 0$, and perfect separation of signal and error can be achieved.

Analogous calculations will hold for the continuous time case: if

$$\hat{y} = \int \gamma_s x_{t-s}\, ds \tag{13}$$

$$\gamma(\omega) = \int \gamma_s e^{-i\omega s}\, ds \tag{14}$$

then

$$\gamma(\omega) = \frac{dF_{yx}(\omega)}{dF_{xx}(\omega)} \tag{15}$$

and $F_{\delta\delta}(\omega)$ is again determined by (7′), provided $dF_{xx} > 0$. If for some frequency $dF_{xx} = 0$, then at that point γ is undetermined, and $dF_{\delta\delta} = 0$. As always in the continuous case (see section 3.4) the integral (13) must be interpreted rather generally, since $\hat{y}$ may involve derivatives of $x(t)$.

5.2 Some Examples of Signal Extraction

If signal $\{y_t)$ and error $\{\eta_t\}$ have rational spectral densities, then evaluation of $\hat{y}$ is straightforward. We shall consider only the simplest examples.

Consider first the case

$$g_{yy} = \frac{\sigma_1^2}{(1 - \alpha z)(1 - \alpha z^{-1})} \tag{1}$$

$$g_{\eta\eta} = \lambda\sigma_1^2, \; g_{y\eta} = 0 \tag{2}$$

Then, by (1.3),

$$\gamma(z) = \frac{\beta}{\lambda\alpha(1 - \beta z)(1 - \beta z^{-1})} \tag{3}$$

where β is the value determined in Ex. (3.3.7). Thus

$$\gamma_s = \frac{\beta^{|s|+1}}{\lambda\alpha(1-\beta^2)} = \frac{\beta^{|s|}}{\Delta} \tag{4}$$

where Δ is also given in Ex. (3.3.7). We have

$$g_{\delta\delta} = \lambda\sigma_1^2\gamma(z) - \frac{\beta\sigma_1^2}{\lambda\alpha(1-\beta z)(1-\beta z^{-1})} \tag{5}$$

so that

$$E(\delta^2) = \frac{\beta\sigma_1^2}{\alpha(1-\beta^2)} = \frac{\lambda\sigma_1^2}{\Delta} \tag{6}$$

If there is little superimposed error, so that λ is small, then β is $O(\lambda)$, Δ is $1 + O(\lambda)$. It is thus plain from (4) and (6) that, as λ tends to zero, y_t tends to x_t itself, and the m.s.e. to zero, as one would expect.

Ex. 1. If $g_{yy} = \sigma^2(1-bz)(1-bz^{-1})$, $g_{\eta\eta} = \lambda\sigma^2$, $g_{y\eta} = 0$, show that

$$\gamma(z) = 1 - \Delta^{-1}\sum_{-\infty}^{\infty}\beta^{|s|}z^s$$

where

$$\Delta = \sqrt{1 + \frac{2(1+b^2)}{\lambda} + \frac{(1-b^2)^2}{\lambda^2}}$$

$$\beta = \frac{2b}{(1+b^2)+\lambda(1+\Delta)}$$

Show also that

$$E(\delta^2) = \lambda\sigma^2\left(\frac{\Delta-1}{\Delta}\right)$$

Note that, as λ becomes small, β tends to the smaller in modulus of b, b^{-1}, and Δ tends to infinity.

The calculations for the analogous continuous time cases follow a similar pattern. There is one interesting point, however; $f_{xx}(\omega)$ may decrease so slowly as to have a divergent integral, indicating that x_t has infinite variance, and yet the l.s. process may yield a perfectly proper estimate with finite m.s.e.

Thus, consider the case analogous to (1), (2):

$$f_{yy} = \frac{\sigma^2}{\omega^2+\alpha^2} \tag{7}$$

$$f_{\eta\eta} = \lambda\sigma^2, f_{y\eta} = 0 \tag{8}$$

We have

$$f_{xx} = f_{yy} + f_{\eta\eta} = \frac{\lambda\sigma^2(\omega^2+\beta^2)}{\omega^2+\alpha^2} \tag{9}$$

where

$$\beta = \sqrt{\alpha^2+\lambda^{-1}} \tag{10}$$

The process x_t has a pure noise component, with the result that $f_{xx}(\omega)$

tends to a non-zero limit at infinity, and is non-integrable. We have, from (1.14) and (1.15),

$$\gamma(\omega) = \frac{1}{\lambda(\omega^2 + \beta^2)} \tag{11}$$

$$\gamma_s = \frac{1}{2\lambda\beta} e^{-\beta|s|} \tag{12}$$

It is readily confirmed that $\hat{y}$ itself has finite variance, and that the m.s.e. is

$$E(\delta^2) = \frac{\sigma^2}{2\beta} = \frac{\sigma^2}{2}\sqrt{\frac{\lambda}{1 + \lambda\alpha^2}} \tag{13}$$

In general, it follows from the fact

$$0 \leqslant \text{var}\,(\hat{y}) + \text{var}\,(\delta) = \text{var}\,(y) \tag{14}$$

that var (δ) and var $(\hat{y})$ will always be finite, whatever the behaviour of x_t, if only the variate being estimated, y, has itself finite variance. Indeed, from (1.7′) we deduce the much stronger relation

$$0 \leqslant F_{\hat{y}\hat{y}}(\omega) + F_{\delta\delta}(\omega) = F_{yy}(\omega) \tag{15}$$

for the case where y is an element of a stationary process.

Ex. 2. Show that if we wish to estimate $\int K_s y_{t-s}\, ds$, where this has finite variance, then the $\gamma(\omega)$ of (1.15) becomes modified to

$$\gamma(\omega) = K(\omega)\frac{dF_{yx}(\omega)}{dF_{xx}(\omega)}$$

Show that this agrees with the conclusion that the l.l.s.e. of a linear function is a linear function of l.l.s.e.'s

Ex. 3. Consider the case of signal extraction in continuous time with $f_{y\eta} = 0$. Show that if we wish to estimate the signal derivative $y'(t)$, then

$$\gamma(\omega) = \frac{i\omega f_{yy}}{f_{yy} + f_{\eta\eta}}$$

Let us now consider the case (7), (8) again with the modification that the error process contains a sinusoidal term of frequency μ. The density $f_{\eta\eta}(\omega)$ will not exist for $\omega = \pm\mu$; at these values there will be a jump in the spectral distribution function of size

$$dF(\mu) = dF(-\mu) = A \tag{16}$$

say. We thus know the amplitude and frequency of the sinusoidal component. In practice, if we knew this much we would probably know the phase as well, and could simply subtract the component out. However, let us suppose that the only information we have is that conveyed in the power spectrum. Then, from (1.15) we find that (11) holds for $\omega \neq \pm\mu$, but that

$$\gamma(\mu) = \gamma(-\mu) = 0 \tag{17}$$

That is, the filter must be chosen just as before, except for the single modification that it shall not pass the frequency μ. If this can be achieved, then the m.s.e. is again given by (13). It cannot be achieved, however, for we determine γ_s from

$$\gamma_s = \frac{1}{2\pi}\int_{-\infty}^{\infty} e^{i\omega s}\gamma(\omega)\, d\omega \tag{18}$$

and the modification (17) will not affect γ_s at all. Conversely, no change in γ_s can bring about the simple modification (17). With γ_s given by (12) the m.s.e. will be

$$E(\delta^2) = \frac{\sigma^2}{2\beta} + |\gamma(\mu)|^2[dF(-\mu) + dF(\mu)] = \frac{\sigma^2}{2\beta} + \frac{2A}{\lambda(\mu^2 + \beta^2)} \tag{19}$$

The difficulty is, that the $\gamma(\omega)$ determined in section (1) may possibly not be representable in terms of its transform, as in (1.14), on a set of measure zero, but that this set will nevertheless be important if it is not also zero in $F_{xx}(\omega)$ measure.

However, one can certainly find a weight-function γ_s which will approach the theoretical optimum performance arbitrarily closely. Suppose we replace (17) by

$$\gamma(\omega) = 0 \qquad \begin{pmatrix} |\omega - \mu| < \Delta \\ |\omega + \mu| < \Delta \end{pmatrix} \tag{20}$$

where Δ is a positive constant. Outside these intervals centred on $\pm\mu$, we shall assume that (11) holds. Then this finite modification of $\gamma(\omega)$ will certainly affect γ_s; we shall have

$$\gamma_s = \frac{1}{2\lambda\beta}e^{-\beta|s|} - \frac{1}{\pi\lambda}\int_{\mu-\Delta}^{\mu+\Delta} \frac{\cos(\omega s)\, d\omega}{\omega^2 + \beta^2} \tag{21}$$

The m.s.e. will be given by

$$E(\delta^2) = \frac{1}{2\pi}\int_{-\infty}^{\infty} dF_{yy}(\omega) - \frac{1}{2\pi}\int_{-\infty}^{\infty} |\gamma(\omega)|^2\, dF_{xx}(\omega) \tag{22}$$

where $\gamma(\omega)$ is given by (11), subject to modification (20). It is thus equal to

$$E(\delta^2) = \frac{\sigma^2}{2\beta} + \frac{\sigma^2}{\lambda\pi}\int_{\mu-\Delta}^{\mu+\Delta} \frac{d\omega}{(\omega^2 + \alpha^2)(\omega^2 + \beta^2)} \tag{23}$$

The modifying term in (23) is of order Δ, and can be made arbitrarily small.

5.3 Interpolation between Evenly-spaced Observations

Consider a process in continuous time, $\{\xi_t\}$, and suppose that ξ is observed at intervals of time τ, yielding a discrete time process

$$x_n = \xi_{n\tau} \qquad (n \text{ integral}) \tag{1}$$

Suppose that on the basis of this sequence we wish to estimate

$$y_n = \xi_{n\tau+\nu} \tag{2}$$

where ν is some value in the range $(0, \tau)$.

Then, if ξ has autocovariance Γ_s and s.d.f. $f(\omega)$, we have, assuming convergence,

$$f_{xx}(\omega) = \sum_n \Gamma_{n\tau} e^{-in\omega} = \frac{1}{\tau}\sum_n f\left(\frac{\omega + 2n\pi}{\tau}\right) \tag{3}$$

The second form for f_{xx} follows from Poisson's summation formula. Similarly,

$$f_{yx}(\omega) = \sum_n \Gamma_{n\tau+\nu} e^{-in\omega} = \frac{1}{\tau}\sum_n f\left(\frac{\omega + 2n\pi}{\tau}\right) e^{i\nu(\omega+2n\pi)/\tau} \tag{4}$$

The optimal set of coefficients γ_j for calculating

$$\hat{y}_n = \hat{\xi}_{n\tau+\nu} = \sum_j \gamma_j x_{n-j} \tag{5}$$

will be determined by inserting expressions (3) and (4) into (1.3).

Thus, to take the simplest example, suppose that

$$f(\omega) = \frac{\sigma^2}{\omega^2 + \alpha^2} \tag{6}$$

$$\Gamma_s = \frac{\sigma^2}{2\alpha} e^{-\alpha|s|} \tag{7}$$

The first expressions in (3) and (4) are obviously the simpler to evaluate: we find

$$f_{xx}(\omega) = \frac{\sigma^2}{2\alpha} \frac{1 - \rho^2}{(1 - \rho z)(1 - \rho z^{-1})} \tag{8}$$

$$f_{yx}(\omega) = \frac{\sigma^2}{2\alpha} \frac{\lambda - \lambda^{-1}\rho^2 + \rho(\lambda^{-1} - \lambda)z^{-1}}{(1 - \rho z)(1 - \rho z^{-1})} \tag{9}$$

where

$$\rho = e^{-\alpha\tau}, \ \lambda = e^{-\alpha\nu} \tag{10}$$

We have thus

$$\gamma(z) = \frac{\lambda - \lambda^{-1}\rho^2 + \rho(\lambda^{-1} - \lambda)z^{-1}}{1 - \rho^2} = A + Bz^{-1} \tag{11}$$

say, and so

$$\hat{\xi}_{n\tau+\nu} = A\xi_{n\tau} + B\xi_{(n+1)\tau} \tag{12}$$

The m.s.e. is readily found to be

$$E(\delta^2) = \frac{\sigma^2}{2\alpha} \frac{(1 - \lambda^2)(\lambda^2 - \rho^2)}{\lambda^2(1 - \rho^2)} \tag{13}$$

One sees that expressions (12) and (13) behave as they should when ν tends to 0 or τ, i.e. λ tends to 1 or ρ.

Ex. 1. Show that if ξ is generated from a pth order stochastic differential equation with random input, so that $f(\omega)$ is of the form (2.5.20), then $\gamma(z)$ is of the form

$$\gamma(z) = \frac{\sum_{-p}^{p-1} d_j z^j}{\sum_{1-p}^{p-1} c_j z^j}$$

where $c_{-j} = c_j$. It is thus not in general true that in this case $\xi_{n\tau+\nu}$ depends only on the $2p$ nearest observations, $\xi_{(n-p+1)\tau}$ to $\xi_{(n+p)\tau}$.

Ex. 2. Use the observations $\xi_{n\tau}$ to estimate

$$y = \frac{1}{N\tau}\int_0^{N\tau} \xi_t \, dt$$

for the case (6). Compare the m.s.e. of $\hat{y}$ with that of $\frac{1}{N+1}\sum_0^N \xi_{n\tau}$.

Ex. 3. Consider the estimation of $\xi_{n\tau+\nu}$ from $\{\xi_{n\tau} + \eta_n\}$, where the η_n are uncorrelated among themselves or with $\{\xi_t\}$.

A final remark; if spectral densities do not exist, then relations (3) and (4) must be expressed in terms of increments of $F(\omega)$ rather than differentials. For instance, (3) becomes

$$dF_{xx}(\omega) = \sum_n dF\left(\frac{\omega + 2n\pi}{\tau}\right) \tag{14}$$

5.4 Smoothing in Several Dimensions

The treatment of section (1) has an immediate formal generalisation for stationary processes in several dimensions, and this generalisation is one of real practical interest.

For example, least-square methods have been used by Thompson (1956) for the smoothing of meteorological maps. The meteorological variables observed over an area will be subject to local variations, which obscure the underlying general weather pattern. These variations are normally smoothed by the freehand drawing of isobars, etc.; Thompson uses the more explicit least-square method. A difficulty is the specification of the spectra of "signal" and "error"; to deduce plausible forms for these would be a research project in itself, but a worthwhile one. Another difficulty is that observations are recorded only at certain points on the plane, so that the problem is strictly not only one of smoothing but also of interpolation between values recorded at points which are irregularly distributed over the plane.

Another application, as yet little explored, would be to geophysical prospecting. Often observations (of, for instance, temperature, temperature gradient, soil type or gravity) are taken at the points of a rectangular lattice on the surface of the ground. L.l.s. methods should be useful for interpolation of these point observations on a "field", or even for extrapolation of the field in the direction of increasing depth. (An extrapolation which is necessary if one is attempting to locate deep heat sources, or gross irregularities of structure or density.)

Such sets of observations can be taken only over a finite lattice, of course, so that we have not the infinite sample postulated. However, in the interior of the lattice edge-effects may be negligible, so that the l.l.s. estimate may scarcely differ from that based upon the assumption of an infinite lattice. The essential fact is that the estimate of the field at a point need not be based only on observations at points lying in a particular relative direction (corresponding to the "past" to which one is so often restricted in temporal processes).

The formulae of section (1) are generalised simply by replacing simple spectral densities by spectral densities with vector arguments. Thus, if we observe a field made up of two mutually uncorrelated components

$$x_{\boldsymbol{\tau}} = y_{\boldsymbol{\tau}} + \eta_{\boldsymbol{\tau}} \tag{1}$$

and wish to estimate the $y_{\boldsymbol{\tau}}$ component:

$$\hat{y}_{\boldsymbol{\tau}} = \int \gamma_{\mathbf{s}} x_{\boldsymbol{\tau}-\mathbf{s}}\, d\mathbf{s} \tag{2}$$

then the weight function $\gamma_{\mathbf{s}}$ will be determined as before by

$$\gamma(\boldsymbol{\omega}) = \int e^{-i\boldsymbol{\omega}\cdot\mathbf{s}} \gamma_{\mathbf{s}}\, d\mathbf{s} = \frac{f_{yy}(\boldsymbol{\omega})}{f_{yy}(\boldsymbol{\omega}) + f_{\eta\eta}(\boldsymbol{\omega})} \tag{3}$$

We have used $\boldsymbol{\tau} = (\tau_1, \tau_2, \ldots \tau_n)$ to describe the co-ordinate of a point. To take what is probably the simplest non-degenerate example, (see Ex. 1 below) suppose that

$$f_{xx}(\boldsymbol{\omega}) = \frac{\phi(\omega)}{\omega^2 + \alpha^2} \tag{4}$$

$$f_{\eta\eta}(\boldsymbol{\omega}) = \lambda\phi(\omega) \tag{5}$$

where $\omega = \sqrt{\omega_1{}^2 + \omega_2{}^2 + \ldots + \omega_n{}^2}$, and α is a real positive constant. We shall regard $\phi(\omega)$ as the spectral density of something approaching pure random variation: that is, ϕ will be fairly flat, but must tend to zero sufficiently quickly at infinity that

$$\int \phi\, d\boldsymbol{\omega} < \infty \tag{6}$$

if one is to avoid the embarrassment of infinite variances.

Then

$$\gamma(\boldsymbol{\omega}) = \frac{1}{1 + \lambda(\omega^2 + \alpha^2)} = \frac{1}{\lambda(\omega^2 + \theta^2)} \tag{7}$$

where

$$\theta = \sqrt{\alpha^2 + \lambda^{-1}} \tag{8}$$

so that the weighting function is given by

$$\begin{aligned} \gamma_{\mathbf{s}} &= \frac{1}{(2\pi)^n} \int e^{-i\boldsymbol{\omega}\cdot\mathbf{s}} \gamma(\boldsymbol{\omega})\, d\boldsymbol{\omega} \\ &= \frac{1}{\lambda(2\pi)^{n/2}} \left(\frac{\theta}{s}\right)^{n/2-1} K_{n/2-1}(\theta s) \end{aligned}$$

The case $n = 2$ will be that of commonest interest.

In his smoothing of meteorological maps Thompson considered a two-dimensional process, and assumed that

$$\Gamma_{\mathbf{s}}^{(yy)} = \sigma^2 e^{-b^2 s^2} \tag{9}$$

$$\Gamma_{\mathbf{s}}^{(\eta\eta)} = k^2 \sigma^2 e^{-a^2 s^2} \tag{10}$$

These functions were not derived from any model, but were chosen as as being convenient and plausible. One has then

$$\frac{f_{\eta\eta}}{f_{yy}} = \left(\frac{kb}{a}\right)^2 e^{-\frac{1}{4}\omega^2\left(\frac{1}{a^2} - \frac{1}{b^2}\right)} = e^{p^2(\omega^2 - \alpha^2)} \tag{11}$$

say. Upon inserting this in (3) we obtain an expression for $\gamma(\boldsymbol{\omega})$ whose Fourier transform $\gamma_{\mathbf{s}}$ is not elementary. Thompson states, without derivation, that for $k \ll 1$, $a \gg b$ (i.e. an error of relatively low variance and auto-correlation) an approximate inversion is given by

$$\gamma_{\mathbf{s}} \sim \frac{\alpha^2}{4\pi} [J_0(\alpha s) + J_2(\alpha s)] \tag{12}$$

If $a = b$, so that the correlation patterns of "signal" and "error" are the same, then $\gamma(\boldsymbol{\omega})$ is constant, and

$$\hat{y}_{\boldsymbol{\tau}} = \frac{1}{1 + k^2} x_{\boldsymbol{\tau}} \tag{13}$$

quite simply, this being in a sense the case in which the l.s. extraction of the y component is least efficient. If p^2 is positive but not large, and $kb/a \gg 1$, then

$$\begin{aligned} \gamma_{\mathbf{s}} &\sim \left(\frac{a}{2\pi kb}\right)^2 \int e^{-i\boldsymbol{\omega}\cdot\mathbf{s} - p^2\omega^2}\, d\boldsymbol{\omega} \\ &= \frac{a^4}{\pi k^2(a^2 - b^2)}\, e^{-s^2/4p^2} \end{aligned} \tag{14}$$

Ex. 1. Let y and ε be random functions of a time-coordinate t and a space co-ordinate $\boldsymbol{\tau} = (\tau_1, \tau_2, \ldots \tau_n)$. Suppose that the process ε is uncorrelated in time, but that in space it has a spectral density function $\phi(\boldsymbol{\omega})$. Suppose also that

$$\frac{\partial y}{\partial t} + \alpha^2 y = \sum_1^n \frac{\partial^2}{\partial \tau_j^2} y + \varepsilon$$

Show that y, regarded purely as a spatial process for a fixed value of t, has s.d.f. proportional to expression (4).

CHAPTER 6

PROJECTION ON THE SEMI-INFINITE SAMPLE

Suppose that we have a "semi-infinite sample" (x_s, $s \leqslant t$) from a stationary process $\{x_t\}$, and that this is to be used to construct the l.l.s.e. $\hat{y}$ of a variate y. In particular, y may be the value of y_t, where $\{y_t\}$ is a process jointly stationary with $\{x_t\}$.

We have already solved the problem of the construction of $\hat{y}$ in section (3.7), and have studied in detail the case of *pure prediction*, where $y_t = x_{t+\nu}$. In this chapter we shall consider a number of more general examples, and shall also re-derive the solution of section (3.7) by a powerful general argument: the *Wiener–Hopf technique*. This, not unnaturally, was the method used by Wiener himself when solving the prediction problem.

The minimum pre-requisite is Ch. 2; some knowledge of Chs. 3 and 5 would be helpful.

6.1. The Prediction Formula

For the reasons given in section (3.1) we restrict our attention to the case of purely non-deterministic processes.

Suppose the l.l.s.e. of y is

$$\hat{y} = \sum \gamma_j x_{t-j} \tag{1}$$

We showed in section (5.1) that if the summation in (1) is unrestricted and if spectral densities exist, then

$$\gamma(z) = \sum \gamma_j z^j = \frac{g_{yx}(z)}{g_{xx}(z)} \tag{2}$$

If summation is restricted in (1) to non-negative values of j, so that y must be predicted from the "past" of the x_t series, then we saw from section (3.7) that

$$\gamma(z) = \frac{1}{\sigma^2 B(z)} \left[\frac{g_{yx}(z)}{B(z^{-1})} \right]_+ \tag{3}$$

where $B(z)$ is a factor of $g_{xx}(z)$ corresponding to the canonical m.a. representation of $\{x_t\}$ [see formulae (3.7.3), (3.7.4)].

In this section we shall re-derive formula (3) by Wiener's method. This is a technique which is less intimately adapted to the problem than the Kolmogorov method of section (3.7); it is less easily generalised to the relatively irregular cases, and it does not indicate so immediately the convergence of the various series. On the other hand, it has the merit of power; it can be used to obtain a quick formal answer in a wide variety of situations (see Ch. 10), questions of convergence and statistical interpretation being cleared up later. Also, comparison of the two is instructive: Kolmogorov argues in the time domain, Wiener in the frequency domain.

We shall prove (3) under the assumption that the series $g_{xx}(z)$, $g_{yx}(z)$ are analytic in a region $\rho \leqslant |z| \leqslant \rho^{-1}$ $(0 < \rho < 1)$, and that $g_{xx}(z)$ is non-zero in this region. These conditions can be considerably relaxed (cf. section (3.8)), but they are fulfilled in most practical cases, and allow a simple proof.

Ex. 1. Show from (3) that if y may be estimated from $(x_s; s \leqslant t + \nu)$, then

$$\gamma(z) = \frac{1}{\sigma^2 B(z)} \left[\frac{g_{yx}(z)}{B(z^{-1})}\right]_{-\nu}$$

The proof of (3) by the Wiener–Hopf method goes formally as follows. Minimising $E[\hat{y} - y]^2$ with respect to the coefficients γ_j, we obtain the equations

$$\sum_{k=0}^{\infty} \Gamma_{j-k}^{(xx)} \gamma_k = \Gamma_j^{(yx)} \qquad (j = 0, 1, 2 \ldots) \tag{4}$$

Multiplying relation (4) by z^j, and adding over *all* integral j we find that

$$g_{xx}(z)\gamma(z) = g_{yx}(z) + h(z) \tag{5}$$

where $h(z)$ is an unknown series in negative powers of z. All series are assumed to be convergent in some annulus containing the unit circle. Dividing (5) by $B(z^{-1})$, we see that

$$\sigma^2 B(z)\gamma(z) = \frac{g_{yx}(z)}{B(z^{-1})} + \frac{h(z)}{B(z^{-1})} \tag{6}$$

Now, the left-hand member of (6) contains no negative powers of z; the second term in the right-hand member contains only negative powers. Identifying coefficients, we thus obtain relation (3).

To justify the argument, we must show that the series in (5) are valid Laurent expansions in an appropriate z region.

First, we show that $\sum \gamma_j^2 < \infty$, so that $\gamma(z)$ converges in $|z| < 1$, and converges at least in the mean on $|z| = 1$.

The expressions

$$\text{var}\,(\hat{y}) = \frac{1}{2\pi} \int_{-\pi}^{\pi} |\sum \gamma_j e^{-i\omega j}|^2 f_{xx}(\omega) d\omega \tag{7}$$

$$\sum \gamma_j^2 = \frac{1}{2\pi} \int_{-\pi}^{\pi} |\sum \gamma_j e^{-i\omega j}|^2 \, d\omega \tag{8}$$

are finite or infinite together, because $f_{xx}(\omega)$ is bounded away from 0 and ∞, by our assumptions on g_{xx}. But

$$\text{var}\,(\hat{y}) \leqslant \text{var}\,(y) < \infty \tag{9}$$

so our assertion is proved.

Now, because $g_{xx}(z)$ is analytic in $\rho \leqslant |z| \leqslant \rho^{-1}$, we must have

$$|\Gamma_j^{(xx)}| \leqslant K\rho^{|j|} \tag{10}$$

where K is a finite constant. Thus, for negative j

$$|\sum_{k=0}^{\infty} \Gamma_{j-k}^{(xx)} \gamma_k| \leqslant K\rho^{-j}\sum_k |\gamma_k|\rho^k = K'\rho^{-j} \tag{11}$$

say, so that $\sum_j z^j \sum_k \Gamma_{j-k}^{(xx)}\gamma_k$ converges at least in $\rho < |z| < \rho^{-1}$, and is validly represented by either of the two members of (5), at least in $\rho < |z| < 1$. This validates the argument, and in fact shows that $\gamma(z)$ is analytic in $|z| < \rho^{-1}$.

Ex. 2. Prediction in the presence of noise. Show that if

$$\left.\begin{aligned} x_t &= u_t + \eta_t \\ y_t &= u_{t+\nu} \end{aligned}\right\} \tag{12}$$

and the "signal" and "noise" processes $\{u_t\}$, $\{\eta_t\}$ are mutually uncorrelated, then

$$\gamma(z) = \frac{1}{\sigma^2 B(z)}\left[\frac{g_{uu}(z)}{z^\nu B(z^{-1})}\right]_+ \tag{13}$$

Ex. 3. Interpolation. Adapt the Wiener–Hopf argument to show that the l.l.s.e. of x_t from $(x_s; s \neq t)$ is determined by

$$\gamma(z) = 1 - \frac{\lambda}{g(z)} \tag{14}$$

where $f(\omega) = g(e^{-i\omega})$ is the s.d.f. of $\{x_t\}$, and

$$\lambda = E(\hat{x}_t - x_t)^2 = \left[\frac{1}{2\pi}\int_{-\pi}^{\pi}\frac{d\omega}{f(\omega)}\right]^{-1} \tag{15}$$

It is interesting to note that the unconditioned, prediction and interpolation variances of x_t, which might be written $\text{var}(x_t)$, $\text{var}(x_t|x_s; s < t)$ and $\text{var}(x_t|x_s; s \neq t)$, are respectively equal to the arithmetic, geometric and harmonic means of the s.d.f. $f(\omega)$.

6.2 The Mean Square Error

We know from the treatment of section (3.6) that

$$E(\delta^2) = \text{var}(y) - \sum_0^\infty c_j^2 = \text{var}(y) - \mathscr{A}|[\phi(z)]_+|^2 \tag{1}$$

where

$$\phi(z) = \sum_{-\infty}^{\infty} c_j z^j = \frac{g_{yx}(z)}{\sigma B(z^{-1})} \tag{2}$$

Ex. 1. Show that for Ex. (1.2)

$$E(\hat{u}_{t+\nu} - u_{t+\nu})^2 = \text{var}(u) - \mathscr{A}\left|\left[\frac{g_{uu}}{\sigma\bar{B}}\right]_\nu\right|^2 \tag{3}$$

Another form is the following:

$$\begin{aligned} E(\delta^2) &= \text{var}\, y - \mathscr{A}|\phi|^2 + \sum_{-\infty}^{-1} c_j^2 \\ &= \text{var}(y) - \mathscr{A}\left[\frac{|g_{yx}|^2}{g_{xx}}\right] + \mathscr{A}\left|\left[\frac{g_{yx}}{\sigma\bar{B}}\right]_-\right|^2 \end{aligned} \tag{4}$$

This expression breaks $E(\delta^2)$ into two components: the first being the m.s.e. for the case when x is calculated from the infinite sample, the second the increase in m.s.e., $\sum_{-\infty}^{-1} c_j^2$, due to the fact that only the "past" of x_t can be used.

6.3 An Example

Consider Ex. (1.2) for the particular case

$$g_{uu}(z) = \frac{\sigma_1^2}{(1-\alpha z)(1-\alpha z^{-1})} \qquad (|\alpha| < 1) \tag{1}$$

$$g_{\eta\eta}(z) = \lambda\sigma_1^2 \tag{2}$$

We have then

$$g_{xx}(z) = \sigma^2 \frac{(1-\beta z)(1-\beta z^{-1})}{(1-\alpha z)(1-\alpha z^{-1})} \tag{3}$$

where β and σ^2 are determined in Ex. (3.3.7). Thus

$$B(z) = \frac{1-\beta z}{1-\alpha z} \tag{4}$$

and

$$\left[\frac{g_{uu}}{z^\nu \bar{B}}\right]_+ = \left[\frac{\sigma_1^2}{z^\nu(1-\alpha z)(1-\beta z^{-1})}\right]_+ = \begin{cases} \dfrac{\sigma_1^2\alpha^\nu}{(1-\alpha\beta)(1-\alpha z)} & (\nu \geqslant 0) \\ \dfrac{\sigma_1^2}{1-\alpha\beta}\left[\dfrac{z^{-\nu}}{1-\alpha z} + \dfrac{\beta(z^{-\nu}-\beta^{-\nu})}{z-\beta}\right] & (\nu \leqslant 0) \end{cases} \tag{5}$$

For the case $\nu \geqslant 0$ (which, at least for $\nu > 0$, can be characterised as prediction of a noise-corrupted signal) we have then

$$\gamma(z) = \frac{\sigma_1^2\alpha^\nu}{\sigma^2(1-\alpha\beta)(1-\beta z)} = \frac{(1-\alpha^{-1}\beta)\alpha^\nu}{1-\beta z} \tag{6}$$

From formula (2.3) we find that the m.s.e. is

$$E(\delta^2) = \frac{\sigma_1^2}{1-\alpha^2} - \frac{\sigma_1^4\alpha^{2\nu}}{\sigma^2(1-\alpha\beta)^2(1-\alpha^2)} = \frac{\sigma_1^2}{1-\alpha^2}\left[1 - \left(\frac{1-\alpha^{-1}\beta}{1-\alpha\beta}\right)\alpha^{2\nu}\right] \tag{7}$$

For the case $\nu = -\mu \leqslant 0$ (signal extraction after lag μ) we find

$$\gamma(z) = \frac{1-\alpha^{-1}\beta}{1-\beta z}\left[z^\mu + \beta\frac{(z^\mu - \beta^\mu)(1-\alpha z)}{z-\beta}\right] \tag{8}$$

Ex. 1. Show from (8) that

$$\gamma_j = \frac{1-\alpha^{-1}\beta}{1-\beta^2}\,[(1-\alpha\beta)\beta^{|j-\mu|} + (\alpha-\beta)\beta^{j+\mu+1}] \tag{9}$$

The m.s.e. is easiest calculated from formula (2.4). We know the value of the first component of (2.4) from formula (5.2.6); the second is equal to

$$\mathscr{A}\left|\left[\frac{g_{uu}}{\sigma\bar{B}}\right]^{(-\mu)}\right|^2 = \frac{\sigma_1^2}{\sigma^2}\frac{\beta^{2\mu+2}}{(1-\alpha\beta)^2(1-\beta^2)}$$
$$= \sigma_1^2\frac{(1-\alpha^{-1}\beta)\beta^{2\mu+2}}{(1-\alpha\beta)(1-\beta^2)} \tag{10}$$

Adding the two components, we have

$$E(\delta^2) = \frac{\sigma_1^2}{1-\beta^2}\left[\alpha^{-1}\beta + \left(\frac{1-\alpha^{-1}\beta}{1-\alpha\beta}\right)\beta^{2\mu+2}\right] \tag{11}$$

Ex. 2. Show that the solutions for $\gamma(z)$ and $E(\delta^2)$ tend to those given in formulae (3.3.4) and (3.3.5) as $\lambda \longrightarrow 0$, if $\nu > 0$, while for $\nu \leqslant 0$ they tend to $z^{-\nu}$ and 0 respectively. Show that if $\nu \longrightarrow -\infty$, then the solutions become equivalent to those given in formulae (5.2.3), (5.2.6).

6.4 Continuous Time Processes

The formal analogue of the treatment of sections (1) and (2) for processes in continuous time is fairly clear: it will certainly be valid if $f_{xx}(\omega)$ and $f_{yx}(\omega)$ are analytic functions of $\omega = \xi + i\eta$; $\log f_{xx}(\omega)$ and $f_{yx}(\omega)$ being regular in some strip $\eta_- \leqslant \eta \leqslant \eta_+$ including $\eta = 0$, and an inequality

$$|\log f_{xx}(\xi + i\eta)| < C|\xi|^{-p} \qquad (p > 0)$$

holding for sufficiently large ξ uniformly for all η in the strip $\eta_- + \varepsilon \leqslant \eta \leqslant \eta_+ - \varepsilon$, $(\varepsilon > 0)$. (See Noble (1958), p. 15.)

Corresponding to (1.3) we shall have

$$\gamma(\omega) = \frac{1}{B(\omega)}\left[\frac{f_{yx}(\omega)}{B(-\omega)}\right]_+ \tag{1}$$

where $B(\omega)$ is determined by the canonical factorisation

$$f_{xx}(\omega) = B(\omega)B(-\omega) \tag{2}$$

In the special case (1.12) this will become

$$\gamma(\omega) = \frac{1}{B(\omega)}\left[\frac{e^{i\nu\omega}f_{uu}(\omega)}{B(-\omega)}\right]_+ \tag{3}$$

If

$$\phi(\omega) = \frac{f_{yx}(\omega)}{B(-\omega)} = \int_{-\infty}^{\infty}\phi_s e^{-i\omega s}\,ds \tag{4}$$

then, corresponding to (2.1) and (2.4),

$$E(\delta^2) = \operatorname{var}(y) - \int_{0-}^{\infty}\phi_s^2\,ds$$
$$= \operatorname{var}(y) - \mathscr{A}\left[\frac{|f_{yx}(\omega)|^2}{f_{xx}(\omega)}\right] + \int_{-\infty}^{0-}\phi_s^2\,ds \tag{5}$$

Again, for the particular case (1.12) this becomes

$$E(\delta^2) = \text{var}\, u - \int_{\nu-}^{\infty} \psi_s{}^2\, ds$$
$$= \mathscr{A}\left[\frac{f_{uu}f_{\eta\eta}}{f_{uu}+f_{\eta\eta}}\right] + \int_{-\infty}^{\nu-} \psi_s{}^2\, ds \tag{6}$$

where

$$\psi(\omega) = \frac{f_{uu}(\omega)}{B(-\omega)} = \int_{-\infty}^{\infty} \psi_s e^{-i\omega s}\, ds \tag{7}$$

The details of the calculations should be fairly clear by now, and we leave the treatment of particular cases as exercises.

Ex. 1. Consider the prediction problem (1.12) in the particular case

$$f_{uu} = \frac{\sigma^2}{\omega^2 + \alpha^2}$$
$$f_{\eta\eta} = \lambda\sigma^2, \quad f_{u\eta} = 0$$

Show that if $\beta = \sqrt{\alpha^2 + \lambda^{-1}}$ then

$$\gamma(\omega) = \begin{cases} \dfrac{(\beta-\alpha)e^{-\alpha\nu}}{\beta + i\omega} & (\nu \geqslant 0) \\[2ex] \dfrac{\beta-\alpha}{\beta+i\omega}\left[e^{i\nu\omega} + \dfrac{\alpha+i\omega}{\beta-i\omega}(e^{i\nu\omega} - e^{\beta\nu})\right] & (\nu \leqslant 0) \end{cases}$$

and that

$$E(\delta^2) = \begin{cases} \dfrac{\sigma^2}{2\alpha}\left[1 - \left(\dfrac{\beta-\alpha}{\beta+\alpha}\right)e^{-2\alpha\nu}\right] & (\nu \geqslant 0) \\[2ex] \dfrac{\sigma^2}{2\beta}\left[1 + \left(\dfrac{\beta-\alpha}{\beta+\alpha}\right)e^{-2\beta\nu}\right] & (\nu \leqslant 0) \end{cases}$$

Ex. 2. If in Ex. 1 we make the modification

$$f_{uu} = \frac{\sigma^2(\mu^2+\omega^2)}{(\alpha_1{}^2+\omega^2)(\alpha_2{}^2+\omega^2)} \qquad [Re(\alpha_1, \alpha_2) > 0]$$

show that

$$\gamma(\omega) = \frac{[(\beta_1-\alpha_1)(\beta_2-\alpha_1)(\alpha_2+i\omega)e^{-\alpha_1\nu} - (\beta_1-\alpha_2)(\beta_2-\alpha_2)(\alpha_1+i\omega)e^{-\alpha_2\nu}]}{(\alpha_2-\alpha_1)(\beta_1+i\omega)(\beta_2+i\omega)}$$

where β_1 and β_2 are the zeros of $\lambda(\alpha_1{}^2+\omega^2)(\alpha_2{}^2+\omega^2) + (\mu^2+\omega^2)$ which have positive real part.

CHAPTER 7

PROJECTION ON THE FINITE SAMPLE

Quite often a record of observations stretching into the distant past is not available, and prediction or estimation must be based on a finite sample; on a series of observations taken over a finite interval of time. This is the case in anti-aircraft gunnery, for example, where the position of the plane must be fixed from observations extending over only a few seconds (although it may well be that observations from earlier instants of time would not materially improve the prediction). In the discrete time case we have then the problem of predicting from a sample $(x_0, x_1, \ldots x_{n-1})$, say, and we know from Ch. 4 that this amounts essentially to the calculation of $\mathbf{V}^{-1}$, where $\mathbf{V}$ is the covariance matrix of the sample variates. In this chapter we consider the calculation of $\mathbf{V}^{-1}$ for variates from certain special stationary processes: the problem is one which has not yet been solved in general. The inversion of $\mathbf{V}$ will also be useful in the next chapter, when we come to consider the fitting of regression sequences.

The prerequisites are Ch. 2 and section (1) of Ch. 4. Comparison will be made with results from Chs. 3, 5 and 6.

7.1 Autoregressive Processes

Suppose we wish to predict $x_{n+\nu-1}$ from $x_0, x_1, \ldots x_{n-1}$. We saw from section (3.2) that if x_t was generated by an autoregression of order p, then $\hat{x}_{n+\nu-1}$ in terms of $(x_s; s < n)$ involved only $x_{n-p}, x_{n-p+1}, \ldots x_{n-1}$. It is clear, then, that, if $p \leqslant n$ the predictor based on the finite sample will be just the predictor derived in Ch. 3. One begins to hopefully expect that it may be equally easy to predict quantities other than $x_{n+\nu-1}$ in this case.

Suppose, then, that we wish to solve the equation system

$$\sum_{k=0}^{n-1} \Gamma_{j-k}\lambda_k = \mu_j \qquad (j = 0, 1, \ldots n-1) \tag{1}$$

for $\lambda_0, \lambda_1, \ldots \lambda_{n-1}$, where

$$g(z) = \sum_{-\infty}^{\infty} \Gamma_s z^s = \frac{1}{A(z)A(z^{-1})} \tag{2}$$

$$A(z) = \sum_{0}^{p} a_j z^j \tag{3}$$

We suppose that $A(z)$ corresponds to an autoregressive operator, and so has all its roots outside the unit circle, also that $p \leqslant n$.

Now, we can replace (1) by an infinite equation system in which j ranges over all integers, and λ_j is regarded as unknown, μ_j known, in the range $0 \leqslant j \leqslant n-1$, while outside this range μ_j is unknown and must be determined, and λ_j is known (equal to zero). It is clear that, as $|j|$ tends to infinity, μ_j tends to zero at the same rate as Γ_j, i.e. as $\rho^{|j|}$ where ρ^{-1} is the smallest zero (in modulus) of $A(z)$. Hence, the generating functions

$$\left.\begin{aligned} \lambda(z) &= \sum_0^{n-1} \lambda_j z^j \\ \mu(z) &= \sum_0^{n-1} \mu_j z^j \\ \mu_1(z) &= \sum_{-\infty}^{-1} \mu_j z^j \\ \mu_2(z) &= \sum_n^{\infty} \mu_j z^j \end{aligned}\right] \tag{4}$$

all converge in $|\rho| < |z| < |\rho|^{-1}$, and for values of z in this annulus we can rewrite the infinite equation system in terms of the functions (4) as

$$g(z)\lambda(z) = \mu_1(z) + \mu(z) + \mu_2(z) \tag{5}$$

We see from (2) and (5) that

$$[A(z^{-1})]^{-1}\lambda(z) = A(z)[\mu_1(z) + \mu(z) + \mu_2(z)] \tag{6}$$

Picking out the terms in $z^n, z^{n+1}, \ldots$ we have then

$$A(z)\mu_2(z) + [A(z)\mu(z)]_n = 0 \tag{7}$$

so that

$$\mu_2(z) = -\frac{1}{A(z)}[A(z)\mu(z)]_n \tag{8}$$

Similarly

$$\mu_1(z) = -\frac{1}{A(z^{-1})}[A(z^{-1})\mu(z)]^{(-1)} \tag{9}$$

Inserting these expressions for μ_1 and μ_2 into (5), we ultimately obtain the solution for $\lambda(z)$:

$$\lambda(z) = A(z)[A(z^{-1})\mu(z)]_0 - A(z^{-1})[A(z)\mu(z)]_n \tag{10}$$

This determines the solution of the equation system (1), and, since the constants μ_j are quite arbitrary, determines the inverse of the matrix $\mathbf{V} = (\Gamma_{j-k})$. In fact, if we choose $\mu_j = w^j$, then we obtain a double generating function for the elements v^{jk} of $\mathbf{V}^{-1}$. We find from (10) with a little manipulation that this is

$$K(z, w) = \sum_0^{n-1} \sum_0^{n-1} v^{jk} z^j w^k = \frac{A(z)A(w) - (zw)^n A(z^{-1})A(w^{-1})}{1 - zw} \tag{11}$$

Let us emphasise again that this solution will not be valid if $p > n$, because then equation (7) would have to include contributions from $A(z)\mu_1(z)$.

Ex. 1. Show that $\mathbf{V}^{-1}$ is symmetric about both diagonals.
Ex. 2. Show that if $A(z) = 1 + az$ and $n \geqslant 2$, then

$$\mathbf{V}^{-1} = \begin{bmatrix} 1 & a & \cdot & \cdot & \cdots \\ a & 1+a^2 & a & \cdot & \cdots \\ \cdot & a & 1+a^2 & a & \cdots \\ \cdot & \cdot & a & 1+a^2 & \cdots \\ \cdot & \cdot & \cdot & \cdot & \cdots \end{bmatrix}$$

Ex. 3. Show that for $A(z) = 1 + az + bz^2$ and $n \geqslant 4$

$$\mathbf{V}^{-1} = \begin{bmatrix} 1 & a & b & \cdot & \cdots \\ a & 1+a^2 & a+ab & b & \cdots \\ b & a+ab & 1+a^2+b^2 & a+ab & \cdots \\ \cdot & b & a+ab & 1+a^2+b^2 & \cdots \\ \cdot & \cdot & \cdot & \cdot & \cdots \end{bmatrix}$$

From (11) we can quickly verify that the predictor of $x_{n+\nu-1}$ is just that given in Ch. 3, a conclusion which is clear on general grounds.

We have $\lambda_j = \gamma_j$, and

$$\mu_j = \Gamma_{j+\nu} = \frac{1}{2\pi i}\oint w^{j+\nu-1} g(w)\, dw \tag{12}$$

so that, from (11),

$$\begin{aligned} \gamma(z) &= \frac{1}{2\pi i}\oint K(z, w) w^{\nu-1} g(w)\, dw \\ &= \frac{1}{2\pi i}\oint \frac{w^{\nu-1}}{1-wz}\left[\frac{A(z)}{A(w^{-1})} - (zw)^n \frac{A(z^{-1})}{A(w)}\right] dw \end{aligned} \tag{13}$$

We can choose the w integration so that $|zw| < 1$; the second term in the integrand then contributes nothing, and the first gives

$$\gamma(z) = A(z)\left[\frac{1}{z^\nu A(z)}\right]_+ \tag{14}$$

which agrees with formula (3.3.7).

Ex. 4. Suppose that the continuous time process $\{\xi_t\}$ has s.d.f. $\sigma^2(\omega^2 + \alpha^2)^{-1}$, and that $x_j = \xi_j$. Use $x_0, x_1, \ldots x_{n-1}$ to predict

$$\int_s^t \xi_u\, du$$

in the cases $(0 \leqslant s; t \leqslant n-1)$, $(s \leqslant 0; t \geqslant n-1)$.
Ex. 5. Show that if $(p \leqslant j, k \leqslant n-p-1)$ then $v^{jk} = g^{(j-k)}$ where

$$g(z)^{-1} = \Sigma g^{(s)} z^s \quad (|z| = 1).$$

7.2 Moving-average Processes

Consider the equation system (1.1) again, but with the $g(z)$ of (1.2) equal to $B(z)B(z^{-1})$. Here $B(z)$ is a polynomial of order q whose zeros shall, for once, be unrestricted in position.

Then equation (1.5) holds as before, but the series $\mu_1(z)$ and $\mu_2(z)$ are finite; in fact, it is clear from (1.5) that μ_j is zero for $j < -q$, $j > n + q - 1$. We have

$$\lambda(z) = \frac{\mu(z) + \mu_1(z) + \mu_2(z)}{B(z)B(z^{-1})} \tag{1}$$

Since $\lambda(z)$ is a polynomial, it can have no poles, and the zeros of the denominator of expression (1) must also be zeros of the numerator. This fact gives us $2q$ linear equations for the $2q$ unknowns $\mu_{-q} \ldots \mu_{-1}$; $\mu_n \ldots \mu_{n+q-1}$.

So, for the case

$$B(z) = 1 - \beta z \tag{2}$$

expression (1) contains two unknowns, μ_{-1} and μ_n, and these are determined by the equations

$$\left.\begin{aligned} \mu(\beta) + \mu_{-1}\beta^{-1} + \mu_n\beta^n &= 0 \\ \mu(\beta^{-1}) + \mu_{-1}\beta + \mu_n\beta^{-n} &= 0 \end{aligned}\right] \tag{3}$$

Ex. 1. Consider the prediction of x_n in the case (2). (The predictor for $\nu > 1$ will be zero.) In this case $\lambda(z) = \gamma(z)$, $\mu(z) = -\beta$. Show that

$$\gamma(z) = (1 - \beta z)^{-1}(1 - \beta z^{-1})^{-1}\left[-\frac{\Delta_n}{\Delta_{n+1}} + z - \frac{\Delta_1 z^{n+1}}{\Delta_{n+1}}\right]$$

$$\gamma_j = -\frac{\Delta_{n-j}}{\Delta_{n+1}} \quad (0 \leqslant j \leqslant n)$$

(cf. formula (3.3.10)), where $\Delta_j = \beta^{-j} - \beta^j$.

Show also (see Ex. (4.1.2)) that

$$E(\delta^2) = \Gamma_0 - \sum_0^{n-1} \gamma_j \Gamma_{j+1} = \sigma^2 \frac{1 - \beta^{2n+4}}{1 - \beta^{2n+2}}$$

We see that the m.s.e. differs from σ^2 only by a term of order β^{2n}, which is, in fact, not likely to be appreciable in many cases.

Ex. 2. Compare the m.s.e. for the optimal predictor of Ex. 1 with that of the "truncated" predictor

$$\hat{x}_{t+\nu} = -\sum_0^{n-1} \beta^{j+1} x_{t-j}.$$

Ex. 3. Show that, for the case (2),

$$\mathrm{v}^{jk} = \frac{\Delta_{j+1}\Delta_{n-k}}{\beta\Delta_1\Delta_{n+1}}$$

if $0 \leqslant j \leqslant k \leqslant n - 1$, while, of course, $\mathrm{v}^{kj} = \mathrm{v}^{jk}$.

Ex. 4. Show from Ex. 3 that if $0 < \nu \leqslant j, k \leqslant n - \nu$ then $\mathrm{v}^{jk} = g^{(j-k)} + O(\beta^\nu)$, where $g^{(s)}$ is defined in Ex. (1.5).

7.3 Processes with Rational S.D.F.

Suppose that

$$g(z) = \frac{B(z)B(z^{-1})}{A(z)A(z^{-1})} \tag{1}$$

where $A(z)$ and $B(z)$ are polynomials of orders p and q respectively. If $p \leqslant n + q$, then the equation system (1.1) can be solved for this case by a combination of the methods of the last two sections.

For brevity we shall denote $A(z)$, $A(z^{-1})$ simply by A, $\bar{A}$, etc.

By imagining that we are solving for $|B|^2\lambda$ rather than for λ, we find as in (1.10) that

$$|B|^2\lambda = A[\bar{A}(\mu + P + Q)]_{-q} - \bar{A}[A(\mu + P + Q)]_{n+q} \tag{2}$$

where P and Q are undetermined finite series of the form

$$\left.\begin{aligned} P(z) &= \sum_{-q}^{-1} P_j z^j \\ Q(z) &= \sum_{n}^{n+q-1} Q_j z^j \end{aligned}\right] \tag{3}$$

With a little manipulation equation (2) can be rewritten

$$|B|^2\lambda = A[\bar{A}\mu]_0 - \bar{A}[A\mu]_n + AP^* + \bar{A}Q^* \tag{4}$$

where P^*, Q^* are polynomials of the same form as P, Q respectively. Now we apply the principles of section (2): the polynomials P^* and Q^* are determined from the fact that all the $2q$ zeros of $|B|^2$ must be zeros of the right-hand member of (4). With this, $\lambda(z)$ is determined.

Ex. 1. Consider the prediction of $x_{n+\nu-1}$, so that $\lambda_j = \gamma_j$, $\mu_j = \Gamma_{j+\nu}$. Show, by generalising the calculation (1.13), (1.14), that if $\nu + n - p - q > 0$, then

$$\gamma(z) = \frac{1}{|B|^2}\left\{A\left[\frac{|B|^2}{z^\nu A}\right]_+ + AP^* + \bar{A}Q^*\right\} \tag{5}$$

Ex. 2. Following on from Ex. 1, show that if

$$A = 1 - \alpha z,\ B = 1 - \beta z$$

then

$$\gamma_j = \frac{\alpha^\nu(1-\alpha\beta)(1-\alpha^{-1}\beta)(\Delta_{n-j} - \alpha\Delta_{n-j-1})}{\beta(\Delta_{n+1} - 2\alpha\Delta_n + \alpha^2\Delta_{n-1})}$$

where $\Delta_j = \beta^{-j} - \beta^j$. Compare with Ex. (3.3.6).

7.4 Interpolation

The problem of interpolation lies somewhat to one side of our main theme, but fits in rather well here. Suppose all terms in a stationary sequence $\{x_t\}$ have been observed, except the block $(x_0, x_1, \ldots x_{n-1})$, and it is desired to estimate these. We have thus a generalisation of Ex. (6.1.3).

If the l.l.s.e. is

$$\hat{x}_\nu = \sum_{\substack{k<0 \\ k\geq n}} \gamma_k x_k \tag{1}$$

then

$$\sum_k \Gamma_{j-k}\gamma_k = \Gamma_{j-\nu} \qquad \begin{pmatrix} j<0 \\ j\geq n \end{pmatrix} \tag{2}$$

Completing the equation system (2) and forming generating functions in the usual way, we have

$$g(z)\gamma(z) = z^\nu g(z) - h(z) \tag{3}$$

where

$$h(z) = \sum_0^{n-1} h_j z^j \tag{4}$$

is an unknown polynomial, to be determined from the conditions $\gamma_0 = \gamma_1 = \ldots = \gamma_{n-1} = 0$. That is,

$$\left[g(z)^{-1}h(z) - z^{\nu}\right]_0^{(n-1)} = 0 \tag{5}$$

so that if $g(z)^{-1}$ has a Laurent expansion on $|z| = 1$,

$$g(z)^{-1} = \Sigma g^{(j)} z^j \tag{6}$$

then

$$\sum_{k=0}^{n-1} g^{(j-k)} h_k = \delta_{j\nu} \qquad (j = 0, 1 \ldots n-1) \tag{7}$$

That is, in order to solve the problem one must solve a finite equation system of the form (1.1), but whose coefficients are now generated by the *reciprocal* of the s.d.f. If $g(z)$ is rational, then the solution can be completed by the methods of sections (1)–(3).

Ex. 1. Moving average case. Suppose $g(z) = B(z)B(z^{-1})$, where $B(z)$ is of degree not greater than n. Show that

$$\sum_n^{\infty} \gamma_j z^j = B^{-1}[z^{\nu}B]_n$$

$$\sum_{-\infty}^{-1} \gamma_j z^j = \bar{B}^{-1}[z^{\nu}\bar{B}]^{(-1)}$$

Ex. 2. Autoregressive case. Suppose that $g(z) = |A(z)|^{-2}$, where the degree p of $A(z)$ is finite, but otherwise unrestricted. Show that $\gamma_j = 0\,(j < -p; j > n + p - 1)$ and that $\gamma(z)$ is determined by

$$\gamma(\xi) = \xi^{\nu}$$

where ξ takes the values of the $2p$ zeros of $A(z)A(z^{-1})$.

Ex. 3. If $A(z) = (1 - \alpha z)/\sigma$, show that

$$\hat{x}_{\nu} = \frac{\Delta_{n-\nu}x_{-1} + \Delta_{\nu+1}x_n}{\Delta_{n+1}}$$

where $\Delta_j = \alpha^{-j} - \alpha^j$, and that

$$E(\hat{x}_{\nu} - x_{\nu})^2 = \frac{\sigma^2}{1 - \alpha^2} \; \frac{\Delta_{\nu+1}\Delta_{n-\nu}}{\Delta_{n+1}}$$

7.5 Continuous Time Processes

We shall restrict attention to processes with rational s.d.f.

$$f(\omega) = \frac{M(i\omega)M(-i\omega)}{L(i\omega)L(-i\omega)} = \left|\frac{M}{L}\right|^2 \tag{1}$$

where $L(\zeta)$, $M(\zeta)$ are polynomials of degrees l and m respectively, with real coefficients, and all zeros in the left half-plane.

Suppose that the l.l.s.e. is to be based on the finite sample $(x_t; 0 \leqslant t \leqslant T)$. We shall have to be able to solve integral equations of the type

$$\int_0^T \Gamma_{t-s}\lambda_s \, ds = \mu_t \qquad (0 \leqslant t \leqslant T) \tag{2}$$

where Γ_s is the autocovariance function of $\{x_t\}$, corresponding to the s.d.f. (1).

Now, just as equation of coefficients of powers of z in (3.2) yields a difference equation for λ_j, so can it be inferred from (1) and (2) that

$$M(D)M(-D)\lambda_t = L(D)L(-D)\mu_t \quad (0 < t < T) \tag{3}$$

where $D = d/dt$. This equation fixes λ_t, to a large extent, in the interior of the region $(0, T)$. However, the integral equation (2) also implies m boundary conditions on λ_t at each point $t = 0$, $t = T$, and also up to l exceptional values for λ_t at each of these points. In fact, λ_t and its derivatives may have grave discontinuities at the end points of the interval, representable in a formal sense by the presence of δ-function components, $\delta(t)$, $\delta(t - T)$, and their derivatives, in λ_t.

These δ-function components are an essential part of the solution, and are not to be regarded as artefacts. However, their presence makes a rigorous discussion of the problem a moderately lengthy and subtle exercise in functional analysis. Since we have not space for this, we shall tackle the problem more formally by setting up the analogues of the generating-function solutions of sections (1)–(3). This will give us all we require.

The method we shall use is in fact relatively easy to justify rigorously: we never write down the "inverse kernel" to Γ_{t-s} explicitly, but rather, generating functions for it. From these generating functions one can evaluate functionals of the type $\int_0^T \xi_s \lambda_s \, ds$ fairly directly, where λ_s is the solution of (2). In general, the ultimate problem is the evaluation of just such a functional.

7.6 Processes Generated by Stochastic Differential Equations

Let $M = 1$ in (5.1), so that we have the continuous time analogue of the autoregression (1.2). Suppose that the solution of the equation (5.2) is formally

$$\lambda_s = \int_0^T k(s, t)\mu_t \, dt \tag{1}$$

In general, $k(s, t)$ will contain δ-function components and their derivatives; for this reason, rather than to solve for k itself, it is better to solve for the *bilinear functional*

$$\mathscr{F}(\xi_s, \eta_t) = \int_0^T \int_0^T k(s, t)\xi_s \eta_t \, ds \, dt \tag{2}$$

In fact, the solutions required are special cases of this functional. For instance, if we wish to predict $x_{T+\nu}$ from $(x_t;\ 0 \leqslant t \leqslant T)$, then it is readily seen that the l.l.s. predictor is

$$\hat{x}_{T+\nu} = \mathscr{F}(x_{T-s}, \Gamma_{\nu+t}) \tag{3}$$

The functional can be written down immediately if ξ_s and η_t are exponentials. For, by analogy with (indeed, as a limiting case of) equation (1.11)

$$K(\theta, \phi) = \int_0^T \int_0^T k(s, t) e^{-\theta s - \phi t}\, ds\, dt$$
$$= \frac{L(\theta)L(\phi) - e^{-(\theta+\phi)T} L(-\theta)L(-\phi)}{\theta + \phi} \tag{4}$$

This expression contains all the information needed to determine $\mathscr{F}$: one has simply to make the correspondences $\theta \to -\dfrac{\partial}{\partial s}$, $\phi \to -\dfrac{\partial}{\partial t}$.
One method of making the calculation explicit is the following: if one separates the terms of even and odd degree in the terms $L(\theta)L(\phi)$, $L(\theta)L(-\phi)$ of (4), then the odd order terms are divisible by $\theta + \phi$, and we can write

$$K(\theta, \phi) = P(\theta, \phi)\left[\frac{1 - E}{\theta + \phi}\right] + Q(\theta, \phi)[1 + E] \tag{5}$$

where

$$E = e^{-(\theta+\phi)T} \tag{6}$$

and P and Q are polynomials, both in fact containing only terms of even degree. Making the correspondence mentioned above, we have then

$$\mathscr{F}(\xi_s, \eta_t) = \int_0^T \left[P\left(-\frac{\partial}{\partial s}, -\frac{\partial}{\partial t}\right)\xi_s \eta_t\right]_{s=t} dt$$
$$+ \left[Q\left(-\frac{\partial}{\partial s}, -\frac{\partial}{\partial t}\right)\xi_s \eta_t\right]_{s=t=0}$$
$$+ \left[Q\left(-\frac{\partial}{\partial s}, -\frac{\partial}{\partial t}\right)\xi_s \eta_t\right]_{s=t=T} \tag{7}$$

As an example, consider the first-order case,

$$L(\zeta) = \zeta + \alpha \tag{8}$$

Then, by (4),

$$K(\theta, \phi) = \frac{(\alpha + \theta)(\alpha + \phi) - E(\alpha - \theta)(\alpha - \phi)}{\theta + \phi}$$
$$= (\alpha^2 + \theta\phi)\left(\frac{1 - E}{\theta + \phi}\right) + \alpha(1 + E) \tag{9}$$

Making the correspondence, as in (7), we have then

$$\mathscr{F}(\xi_s, \eta_t) = \int_0^T [\alpha^2 \xi_t \eta_t + \xi_t' \eta_t']\, dt + \alpha[\xi_0 \eta_0 + \xi_T \eta_T] \tag{10}$$

It will be useful to recast this by partial integration so that one of the variables, say ξ_t, is free of differentials under the integral. We find readily that this alternative form is

$$\mathscr{F}(\xi, \eta) = \int_0^T \xi_t(\alpha^2\eta_t - \eta_t'')\, dt + \xi_0(\alpha\eta_0 - \eta_0') + \xi_T(\alpha\eta_T + \eta_T') \tag{11}$$

So, suppose we are considering prediction in this case, and that

$$f(\omega) = \frac{1}{\omega^2 + \alpha^2} \tag{12}$$

We see from (3) and (11) that

$$\begin{aligned} \hat{x}_{T+\nu} &= \mathscr{F}(x_{T-s}, e^{-\alpha(\nu+t)}/2\alpha) \\ &= e^{-\alpha\nu}x_T \end{aligned} \tag{13}$$

In this case, as for the general case of this section, the prediction formula does not differ from that based on a semi-infinite sample. Nevertheless, the general solutions (4) and (7) will be needed in the Ch. 8, when we consider the fitting of regression functions.

It is interesting to note that expression (13) is derived entirely from the boundary term for $t = 0$ in (11): the integral and the term for $t = T$ vanish. In other words, for this case it is a δ-component of $k(s, t)$, rather than the finite part, which determines the solution completely.

Ex. 1. Suppose that

$$L(\zeta) = \zeta^2 + a_1\zeta + a_2$$

Show that

$$\begin{aligned} \mathscr{F}(\xi, \eta) = \int_0^T \xi_t L(D)L(-D)\eta_t\, dt \\ + [-\xi' L(-D)\eta + \xi(D + a_1)L(-D)\eta]_{t=0} \\ + [\xi' L(D)\eta + \xi(-D + a_1)L(D)\eta]_{t=T} \end{aligned}$$

Hence show that $\hat{x}_{t+\nu}$ has the value which would be given by the treatment of section (3.5).

Ex. 2. Extend to general l.

7.7 Continuous Time Processes with Rational S.D.F.

To solve (5.2) for the general case (5.1) one can employ a method of solution analogous to that implied in equation (3.4). More explicitly, the double generating function defined in (6.4) has now the form

$$K(\theta, \phi) = \frac{1}{M(\theta)M(-\theta)}\left[\frac{L(\theta)L(\phi) - EL(-\theta)L(-\phi)}{\theta + \phi} + L(\theta)P(\theta) + e^{-\theta T}L(-\theta)Q(\theta)\right] \tag{1}$$

where $P(\theta)$, $Q(\theta)$ are polynomials in θ of degree $m - 1$, which are to be chosen so that the zeros of $M(\theta)M(-\theta)$ are also zeros of the bracketed numerator.

As an example, consider the prediction of $x_{T+\nu}$ in the case

$$f(\omega) = \frac{\omega^2 + \beta^2}{(\omega^2 + \alpha_1^2)(\omega^2 + \alpha_2^2)} \tag{2}$$

for which

$$\begin{aligned}\Gamma_{s+\nu} &= \frac{1}{2(\alpha_2^2 - \alpha_1^2)}\left[\frac{\beta^2 - \alpha_1^2}{\alpha_1}e^{-\alpha_1(s+\nu)} - \frac{\beta^2 - \alpha_2^2}{\alpha_2}e^{-\alpha_2(s+\nu)}\right] \qquad (3)\\ &= C_1 e^{-\alpha_1 s} + C_2 e^{-\alpha_2 s}\end{aligned}$$

say. Then

$$\begin{aligned}\gamma(\omega) &= \int_0^T \gamma_s e^{-iws}\,ds\\ &= \int_0^T\int_0^T e^{-iws}k(s, t)\Gamma_{t+\nu}\,ds\,dt\\ &= C_1K(i\omega, \alpha_1) + C_2K(i\omega, \alpha_2) \qquad (4)\end{aligned}$$

Consider the evaluation of $K(i\omega, \alpha_1)$, for example. We have, from (1),

$$\begin{aligned}K(i\omega, \alpha_1) = \frac{1}{\omega^2 + \beta^2}[2\alpha_1(\alpha_1 + \alpha_2)(i\omega + \alpha_2) + P(i\omega + \alpha_1)(i\omega + \alpha_2) +\\ Qe^{-iwT}(-i\omega + \alpha_1)(-i\omega + \alpha_2)] \qquad (5)\end{aligned}$$

where P and Q are constants, to be chosen so that the bracketed numerator is zero when $i\omega = \pm\beta$. In this way we find that

$$P = \psi[(\alpha_1 + \beta)(\alpha_2 + \beta)^2 e^{\beta T} - (\alpha_1 - \beta)(\alpha_2 - \beta)^2 e^{-\beta T}] \tag{6}$$

$$Q = \psi[(\alpha_2^2 - \beta^2)2\beta] \tag{7}$$

where

$$\psi = \frac{2\alpha_1(\alpha_1 + \alpha_2)}{e^{\beta T}L^2(\beta) - e^{-\beta T}L^2(-\beta)} = \frac{2\alpha_1(\alpha_1 + \alpha_2)}{\Delta} \tag{8}$$

say. The transform of expression (5) is

$$\lambda_s = -P\delta(s) - Q\delta(s - T) - \frac{Q}{2\beta}[L(\beta)e^{\beta(T-s)} - L(-\beta)e^{\beta(s-T)}] \tag{9}$$

This can be seen directly, but an easier way is to note that λ_s must be of the form (9), and to calculate the actual coefficients by back-substitution; i.e. by transforming (9) and identifying it with (5).

Combining equations (4) – (9) we find that

$$\hat{x}_{T+\nu} = cx_T + dx_0 + d\int_0^T [L(\beta)e^{\beta(T-s)} - L(-\beta)e^{\beta(s-T)}]x_{T-s}\,ds \tag{10}$$

where

$$c = \frac{e^{-\alpha_1 \nu}}{(\alpha_2 - \alpha_1)\Delta}[L^2(-\beta)(\alpha_1 + \beta)e^{-\beta T} - L^2(\beta)(\alpha_1 - \beta)e^{\beta T}]$$

$$- \frac{e^{-\alpha_2 \nu}}{(\alpha_2 - \alpha_1)\Delta}[L^2(-\beta)(\alpha_2 + \beta)e^{-\beta T} - L^2(\beta)(\alpha_2 - \beta)e^{\beta T}] \quad (11)$$

$$d = \frac{L(\beta)L(-\beta)}{(\alpha_2 - \alpha_1)\Delta}(e^{-\alpha_2 \nu} - e^{-\alpha_1 \nu}) \quad (12)$$

If we let T become infinite, then formula (10) converges to the earlier form (3.6.5).

CHAPTER 8

DEVIATIONS FROM STATIONARITY: TRENDS, DETERMINISTIC COMPONENTS AND ACCUMULATED PROCESSES

One often has to deal with non-stationary series; indeed, except in physical data, it is rare to find a series at all which is truly stationary. For example, population and economic series are evolutive, almost by their very nature. In tracking an approaching enemy aircraft one has a non-stationary situation, and even if one tries to avoid this by postulating a stream of aircraft, there is still the consideration that uniformly good prediction is unnecessary; the only error that matters is the one made at the instant of firing.

In dropping the assumption of stationarity, one is left with scarcely any restriction upon one's model. For this reason, it is all the more difficult to specify a model, or even to specify some of the statistical properties of the variates (such as first and second moments). In consequence, methods of prediction for non-stationary processes have tended to be more or less empiric; i.e. based upon intuitive ideas rather than upon deduction from an explicitly postulated class of models.

In this chapter we consider very briefly the fitting of linear models, and the classical methods of trend fitting. Then we consider in greater detail the l.s. fitting of regression functions, or deterministic components (cf. section (4.3)). We then deal with accumulated processes, i.e, processes which by repeated differencing or differentiation can be made stationary. These have recently been studied by Yaglom (1955); they can be treated as explicitly as stationary processes, offer a rather flexible concept of the idea of "trend" and their prediction formulae are automatically exact for polynomial sequences up to a certain degree. Finally, we consider the so-called method of "exponentially weighted moving averages": this is seen to be a special way of looking at methods already proposed.

The prerequisites are Chs. 2, 3 and 4.

8.1 Linear Models

Suppose the variates x_1, x_2, . . . have known means (which we may subtract, and so suppose zero) and covariances. Then, in principle, one can apply the l.s. methods of Ch. 4 to predict $x_{n+\nu}$ from x_1, x_2, . . . x_n, say. A natural way of doing this is that described in formulae (4.2.5)–(4.2.8): to orthogonalise the x's consecutively, and to represent the prediction in terms of the orthogonalised variates ε_t. Using the notation of section (4.2), we should then have a predictor

$$\hat{x}_{n+\nu} = \sum_{j=1}^{n} d_{n+\nu,j}\varepsilon_j = \sum_{j=1}^{n}\sum_{k=1}^{j} d_{n+\nu,j}c_{jk}x_k \tag{1}$$

with m.s.e.

$$E(\delta^2) = \sum_{j=n+1}^{n+\nu} d^2_{n+\nu,j} \tag{2}$$

However, it is unlikely that one would be able to specify the means and covariances of the x_t: one would rarely have the information *a priori*, and to estimate so many quantities from a sample is impossible.

It is probably desirable to keep in mind the possibility (or necessity) of inference from the sample: one's *a priori* information is often so indefinite that a practical procedure must allow one to estimate any parameters or functions necessary for prediction from the very sample one is using as prediction basis. Thus, the number of parameters must certainly be small relative to the number of observations.

A plausible restriction on the class of models is the following: that if the process itself is not invariant under time-translation, at least the relations generating it may be. In particular, one might attempt to represent x_t as being generated by an autoregression with constant coefficients,

$$\sum_{0}^{p} a_j x_{t-j} = \mu + \varepsilon_t \tag{3}$$

(μ representing the possible non-zero mean) with the difference that the zeros of $A(z) = \sum a_j z^j$ are not restricted to lying outside the unit circle, as they were in the stationary case. Once the coefficients a_j have been determined, then relation (3) can be used for prediction, just as in section (3.3), by applying it repeatedly with the unknown ε's set equal to zero. In general, these coefficients must be estimated from the sample, $x_1, x_2, \ldots x_n$, by minimisation of

$$\sum_{t=p+1}^{n} (\sum_{j} a_j x_{t-j} - \mu)^2 \tag{4}$$

with respect to $a_1, a_2, \ldots a_p$ and μ.

If, as sometimes happens, the variance of ε varies with the current level of x, then one has a rather more complicated situation.

8.2 Extrapolation of Trends

There was a time, some twenty or more years ago, when by "prediction" or "extrapolation" of a time series, one meant simply the determination of a "trend" by the fitting of a polynomial in time to the known series, or to the latter part of it, and then the continuation of this polynomial as far into the unknown future as was consistent with plausibility. To be fair, one should add that all except the rashest exponents of this procedure seem to have regarded it with some degree of healthy mistrust, for it was rarely described without a caution against its uncritical application.

What is interesting, is that the presence of an evolutive term (trend) is taken for granted, and one attempts to predict the evolutive term rather than the stationary variation about it. This is probably a fair attitude, if one is dealing with economic or demographic series, as was commonly the case.

A second point is, that no very explicit model is assumed, except the representation of the trend as a polynomial, and it is the patent crudeness and limited validity of this assumption that makes one suspicious of the procedure. Furthermore, without at least a partial model one cannot assign an estimate of error to a prediction.

Sometimes, indeed, a model was set up, of the type

$$x_t = T_t + S_t + \eta_t \tag{1}$$

where T_t is a trend, often assumed polynomial, S_t is a periodic term accounting for seasonal variation, often represented by a few sinusoids, and η_t is either a purely random sequence, or, in later work, some more general type of stationary sequence.

If one is really convinced that (1), with polynomial T_t and sinusoidal S_t, is a valid model, then one should fit T and S by some l.s. procedure using all available data (see the next section). However, workers have often been rightly suspicious of the polynomial representation of T, and have regarded this as only locally valid, i.e. as an approximation holding only over short sections of the series. For this reason T_t will generally be fitted only to the last few observations, $x_1, x_2, \ldots x_n$, say. If there is no seasonal term, and if one supposes

$$T_t = \sum_{j=0}^{p} a_j t^j \tag{2}$$

where $p + 1 \leqslant n$, then the convention has been to determine the a's by minimising

$$\sum_{t=1}^{n} (x_t - \sum_{0}^{p} a_j t^j)^2 \tag{3}$$

An alternative way of arranging the calculation is to express T_t in terms of the orthogonal polynomials

$$P_{jn}(t) = \sum_{k=0}^{j} c_{jk} t^{j-k} \tag{4}$$

for which $c_{j0} = 1$ and

$$\sum_{t=1}^{n} P_{jn}(t) P_{kn}(t) = 0 \qquad (j \neq k) \tag{5}$$

In terms of these the predicted trend is

$$T_{n+\nu} = \sum_{j=0}^{p} \left[\frac{P_{jn}(n+\nu) \sum_{t=1}^{n} P_{jn}(t) x_t}{\sum_{t=1}^{n} P_{jn}^2(t)} \right] \tag{6}$$

If one is willing to accept assumptions (1) and (2), then an error can be attached to the prediction (6) (see the next section): not otherwise.

8.3 The Fitting of Deterministic Components

Let us take a rather more explicit form of model (2.1):

$$x_t = \sum_{1}^{q} \beta_j g_j(t) + \xi_t \tag{1}$$

where the $g_j(t)$ are known functions of t, and ξ_t a purely non-deterministic process of the type we have considered in Chs. 3 and 5, with known s.d.f. The $g_j(t)$ will in general be made up of sinusoidal terms and powers of t, corresponding to the "seasonal" and "trend" terms of section (2). The model is presumably as much open to criticism as it was in that section. A good deal depends on the intended application, however. Model (1) is undoubtedly too rigid for econometric and demographic series. On the other hand, it has been used to some extent in the prediction of aircraft flight paths, which are fairly smooth, most of the ξ_t variation coming from the tracking equipment rather than from the motion of the aircraft itself. Presumably the model must still be "local" in character, i.e. valid only over short runs, but at least one can hope to gather enough data that adequate estimates can be made of the $\{\xi\}$ s.d.f. and of the probable magnitudes of specification errors.

Suppose that the quantity one wishes to estimate is

$$y = \sum_{1}^{q} \beta_j h_j + \zeta \tag{2}$$

where the h_j's are known constants, and ζ a random variable with zero mean, whose distribution jointly with the ξ_t is known. For example, in the case of pure prediction, when $y = x_{t+\nu}$, one would have $h_j = g_j(t+\nu)$, $\zeta = \xi_{t+\nu}$. A more realistic case is that of disturbed prediction, when

$$\left.\begin{aligned} y &= u_{t+\nu} \\ x_t &= u_t + \eta_t \\ u_t &= \sum_{1}^{q} \beta_j g_j(t) + \zeta_t \end{aligned}\right\} \tag{3}$$

We have placed the deterministic terms $\sum \beta_j g_j(t)$ entirely in the "signal" component $\{u_t\}$ and not at all in the "noise" component $\{\eta_t\}$: this will be realistic in most applications.

Now, if the β_j were known, one would simply subtract the entire deterministic component, and predict ζ from the known ξ_t by the methods already developed. However, the β_j will in general be unknown. They may indeed be random variables, specific in value for a realisation, but varying in value from realisation to realisation (for example, they might be regarded as specifying an aircraft's flight path in an individual run).

In section (4.3) we considered just this problem: the prediction of y from a set of values of x_t when the coefficients β_j are unknown in value. We found that a minimax criterion (4.3.4) led to the demand (4.3.7), that the conventional least-square approach be supplemented by the condition that the estimator be *exact* for the sequences $g_j(t)$. That is, the l.l.s.e. $\hat{y}$ must be so constructed that if

$$x_t = g_j(t) \tag{4}$$

then

$$\hat{y} = h_j \tag{5}$$

$(j = 1, 2, \ldots q)$. This led in turn to the conclusion that the observed x_t should be used to form estimates $\hat{\beta}_j$ of the β_j, the $\hat{\beta}_j$ being identical with the maximum likelihood estimates in the case of normally distributed ξ_t. We have then

$$\hat{y} = \sum \hat{\beta}_j h_j + \hat{\zeta} \tag{6}$$

where $\hat{\eta}$ is predicted from the

$$\hat{\xi}_t = x_t - \sum \hat{\beta}_j g_j(t) \tag{7}$$

in the same way that ζ would have been predicted from the ξ_t, if these could have been observed directly.

We shall assume that the sample of x's is a finite one: $(x_0, x_1, \ldots x_{n-1})$ in the discrete case, $(x_t; 0 \leqslant t \leqslant T)$ in the continuous one. There are three reasons for this: the model (1) is only to be regarded as an approximation, valid over time intervals of some maximum length; in applications such as flight prediction one will have only a finite interval of observation; and if one has infinitely many x_t and the $g_j(t)$ do not grow too quickly, then the problem becomes trivial, in that one can estimate the β_j with zero sampling variance, even with cruder estimates than the $\hat{\beta}_j$.

The only new problem, then, is the calculation of the $\hat{\beta}_j$. We saw in (4.3.4) that the vector of estimates was given by

$$\hat{\boldsymbol{\beta}} = (\mathbf{G}^{\dagger}\mathbf{V}^{-1}\mathbf{G})^{-1}\mathbf{G}^{\dagger}\mathbf{V}^{-1}\mathbf{X} \tag{8}$$

where $\mathbf{X}$ is the vector $(x_0, x_1, \ldots x_{n-1})$, $\mathbf{V}$ the covariance matrix of $\xi_0, \xi_1, \ldots \xi_{n-1}$, and $\mathbf{G}$ is the $n \times q$ matrix with elements $g_j(t)$, $(j = 1, 2 \ldots q; t = 0, 1, \ldots n - 1)$. The calculation of $\mathbf{V}^{-1}$ is the principal difficulty: here we shall use the results of Ch. 7.

It is worth noting that

$$E\hat{\boldsymbol{\beta}} = \boldsymbol{\beta} \tag{9}$$

$$\mathbf{V}_{\beta\beta} = E(\hat{\boldsymbol{\beta}} - \boldsymbol{\beta})(\hat{\boldsymbol{\beta}} - \boldsymbol{\beta})^{\dagger} = (\mathbf{G}^{\dagger}\mathbf{V}^{-1}\mathbf{G})^{-1} \tag{10}$$

When fitting sinusoids it is simpler to regard them as sums of complex exponentials; in this case **G** will be complex, which is why we have $\mathbf{G}^{\dagger}$ rather than $\mathbf{G}'$ in all the above formulae.

8.4. Deterministic Components: Special Cases

Let us denote the vector of values of the jth regression function, $[g_j(0), g_j(1), \ldots g_j(n-1)]$, by $\mathbf{G}_j$, and set

$$r_j = \mathbf{G}_j^{\dagger}\mathbf{V}^{-1}\mathbf{X} \tag{1}$$

$$m_{jk} = \mathbf{G}_j^{\dagger}\mathbf{V}^{-1}\mathbf{G}_k \tag{2}$$

so that (3.8) and (3.10) can be written

$$\hat{\boldsymbol{\beta}} = (m_{jk})^{-1}(r_j) \tag{3}$$

$$\mathbf{V}_{\beta\beta} = (m_{jk})^{-1} \tag{4}$$

Consider now the important special case when the regression components are exponential, $g_j(t) = \theta_j^{\,t}$, and ξ is autoregressive, order p,

$$f_{\xi\xi}(\omega) = \frac{\sigma^2}{A(z)A(z^{-1})} \tag{5}$$

If $p \leqslant n$ then

$$m_{jk} = \sigma^{-2}K(\bar{\theta}_j, \theta_k) \tag{6}$$

where K is given in equation (7.1.11). From (7.1.11) we also find, by identification of coefficients, that

$$r_j = \sum_{t=0}^{n-1} x_t\bar{\theta}_j^{\,t}\Big(\sum_{s=0}^{t} a_s\bar{\theta}_j^{\,-s}\Big)\Big(\sum_{s=0}^{n-t-1} a_s\bar{\theta}_j^{\,s}\Big) \tag{7}$$

Equations (6) and (7) essentially solve the problem in this special case, in that all that is left is the inversion of the $q \times q$ matrix (m_{jk}).

If $|\theta_j| = 1$ then m_{jj} must be evaluated by de l'Hôpital's rule:

$$m_{jj} = \sigma^{-2}[n|A(\theta_j)|^2 - \theta_j A'(\theta_j)A(\bar{\theta}_j) - \bar{\theta}_j A'(\bar{\theta}_j)A(\theta_j)] \tag{8}$$

Ex. 1. Suppose one is fitting a simple mean, so that $x_t = \beta + \xi_t$. Show that if we are in case (5) with $A(z) = 1 - \alpha z$, then

$$\hat{\beta} = \frac{x_0 + x_{n-1} + (1-\alpha)\sum_{1}^{n-2} x_t}{2 + (1-\alpha)(n-2)}$$

$$\operatorname{var}\hat{\beta} = \frac{\sigma^2}{2(1-\alpha) + (n-2)(1-\alpha)^2} \sim \frac{\sigma^2}{n(1-\alpha)^2}$$

Ex. 2. For the example of Ex. 1 consider the "crude" l.s.e.

$$\breve{\beta} = \frac{1}{n}\sum_{0}^{n-1} x_t$$

obtained by minimisation of $\Sigma(x_t - \beta)^2$. Show that

$$\operatorname{var} \breve{\beta} = \frac{\sigma^2[n(1-\alpha^2) - 2\alpha + 4\alpha^{n+1} - 2\alpha^{n+2}]}{n^2(1-\alpha^2)(1-\alpha)^2}$$

$$\sim \frac{\sigma^2}{n(1-\alpha)^2}$$

Thus the asymptotic efficiency of $\breve{\beta}$, $\lim_{n\to\infty} \operatorname{var} \hat{\beta}/\operatorname{var} \breve{\beta}$, is 1.

Ex. 3. If $x_t = \beta\theta^t + \xi_t$, $A(z) = 1 - \alpha z$, show that the asymptotic efficiency of

$$\breve{\beta} = \frac{\Sigma\bar{\theta}^t x_t}{\Sigma|\theta|^{2t}}$$

is $\dfrac{1-\alpha^2}{1-|\alpha\theta|^2}$ if $|\theta| < 1$, $\dfrac{1-\alpha^2}{|\theta|^2 - \alpha^2}$ if $|\theta| > 1$.

Suppose that all the θ_j lie on the unit circle, so that

$$\theta_j = e^{i\omega_j} \tag{9}$$

say. Then from (6), (7) and (8) we see that

$$m_{jj} = \frac{n}{f(\omega_j)} - \frac{\partial}{\partial\omega_j}\frac{1}{f(\omega_j)} = \frac{n}{f(\omega_j)} + O(1) \tag{10}$$

$$m_{jk} = O(1) \qquad\qquad j \neq k \tag{11}$$

$$r_j = \frac{\sum_{t=0}^{n-1} x_t e^{-i\omega_j t}}{f(\omega_j)} + O(1) \tag{12}$$

By the term O(1) in (12) we mean a random variable whose variance is O(1).

In view of (10)–(12), then, we have from (3) and (4) that for large n

$$\hat{\beta}_j \sim \frac{1}{n}\sum_0^{n-1} x_t e^{-i\omega_j t} \tag{13}$$

$$\operatorname{cov}(\hat{\beta}_j, \hat{\beta}_k) \sim \delta_{jk}\frac{f(\omega_j)}{n} \tag{14}$$

In other words, if sinusoids alone are being fitted, then the l.l.s. estimators $\hat{\beta}_j$ are approximately given by expressions (13) (which are also approximately equal to the "crude" l.s. estimators obtained by minimisation of $\sum_0^{n-1} [x_t - \sum\beta_j g_j(t)]^2$), and these estimates are approximately uncorrelated, with variances given by (14). [See Whittle (1952), and for a more general and rigorous treatment, Grenander (1954).]

While these results have been demonstrated, loosely at that, only for autoregressive processes, they will hold asymptotically for any process which can be approximated arbitrarily well, in a suitable sense, by a finite autoregression. Grenander has given exact conditions, and shown that the result certainly holds if $f(\omega)$ is piece-wise continuous, and is bounded away from both zero and infinity.

Ex. 4. By using the results of section (7.2), show that if $x_t = \beta + \xi_t$, and

$$f(\omega) = \sigma^2(1 - bz)(1 - bz^{-1})$$

then

$$\hat{\beta} = \frac{1}{c(1-b)^2}\sum_0^{n-1}\left[1 - \frac{\Delta_{t+1} + \Delta_{n-t}}{\Delta_{n+1}}\right]x_t$$

$$\text{var}\,\hat{\beta} = \frac{\sigma^2}{c}$$

where $\Delta_j = b^{-j} - b^j$, $c = \dfrac{1}{(1-\text{b})^2}\left[n - 2\dfrac{\Delta_{n/2}\Delta_{(n+1)/2}}{\Delta_{1/2}\Delta_{n+1}}\right]$

Ex. 5. For the case of Ex. 4, show that

$$\text{var}\,\breve{\beta} = \sigma^2\frac{n(1-b)^2 + 2b}{n^2}$$

where $\breve{\beta}$ is defined in Ex. 2.

Ex. 6. Hence show that both var $\hat{\beta}$ and var $\breve{\beta}$ tend to $\sigma^2(1-b)^2/n = f(0)/n$ if $b \neq 1$, but that if $b = 1$, so that $f(0) = 0$, then

$$\text{var}\,\hat{\beta} = \frac{12\sigma^2}{n(n+1)(n+2)}$$

$$\text{var}\,\breve{\beta} = \frac{2\sigma^2}{n^2}$$

This shows what may happen when $f(\omega)$ has a zero at one of the frequencies occurring in the deterministic component. On the other hand, although the two variances are of different orders of magnitude in n, it is still true to say that they both differ from $f(0)/n$ only by terms which are $o(n^{-1})$.

These approximations hold only for θ_j on the unit circle; Ex. 3 gives an indication that they will not hold if $|\theta| \neq 1$. However, they do continue to hold for sequences obtained by differentiating the sequence $e^{i\omega_j t}$ w.r.t. ω_j, i.e., for sequences $g(t) = t^\nu e^{i\omega_j t}$, and in particular, if $\omega_j = 0$, for polynomials in t.

Ex. 7. If $x_t = \beta_1 + \beta_2\left(t - \dfrac{n-1}{2}\right) + \xi_t$, $f(\omega) = \sigma^2[(1-\alpha z)(1-\alpha z^{-1})]^{-1}$, show that $\hat{\beta}_1$ has the value and variance of Ex. 1, and that

$$\hat{\beta}_2 = \frac{1}{c}\left[\left(\frac{n-3}{2}\alpha - \frac{n-1}{2}\right)(x_0 - x_{n-1}) + (1-\alpha)^2\sum_2^{n-2}\left(t - \frac{n-1}{2}\right)x_t\right]$$

$$\to \frac{\sum_0^{n-1}\left(t - \frac{n-1}{2}\right)x_t}{\sum_0^{n-1}\left(t - \frac{n-1}{2}\right)^2}$$

$$\text{var}\,\hat{\beta}_2 = \frac{\sigma^2}{c} \to \frac{12\sigma^2}{n(n^2-1)(1-\alpha)^2}$$

where

$$c = \frac{n(n^2-1)(1-\alpha)^2}{12} + \frac{n-1}{2}[(n+1)\alpha - (n-1)\alpha^2]$$

and that $\hat{\beta}_1$, $\hat{\beta}_2$ are uncorrelated.

Exact formulae for the m_{jk} and r_j in this case can evidently be obtained by differentiating expressions (6) and (7) with respect to the θ arguments.

However, we shall merely note the generalisations of approximations (13) and (14): if one is considering the representation

$$x_t = \sum\sum \beta_{jk} e^{i\omega_j t} P_{kn}(t) + \xi_t \tag{15}$$

where the $P_{kn}(t)$ are the orthogonal polynomials of equation (2.4), (but defined on an interval $t = 0, 1 \ldots n - 1$ rather than $t = 1, 2 \ldots n$) then

$$\hat{\beta}_{jk} \sim \frac{\sum_{t=0}^{n-1} e^{-i\omega_j t} P_{kn}(t) x_t}{\sum_{t=0}^{n-1} P_{kn}{}^2(t)} \tag{16}$$

and these estimates are approximately uncorrelated, with variances

$$\text{var}\,(\hat{\beta}_{jk}) \sim \frac{f(\omega_j)}{\sum_{t=0}^{n-1} P_{kn}{}^2(t)} \tag{17}$$

The analogues of (15)–(17) for the case of continuous time are fairly plain. The analogues of the exact results are also direct, and tend to be slightly simpler than the discrete results. Thus, if $g_j(t) = e^{-\lambda_j t}$ $(j = 1, 2, \ldots q)$ and

$$f(\omega) = \frac{\sigma^2}{L(i\omega)L(-i\omega)} \tag{18}$$

where L is the polynomial of section (7.5), then corresponding to (6)

$$m_{jk} = \sigma^{-2} K(\bar{\lambda}_j, \lambda_k) \tag{19}$$

where K is now determined by equation (7.6.4). Similarly, the Laplace transform of the function weighting x_t in r_j is $K(\bar{\lambda}_j, \theta)$.

Ex. 8. If $x_t = \beta + \xi_t$, $L(\theta) = \theta + \alpha$, show that

$$\hat{\beta} = \frac{x_0 + x_T + \alpha \int_0^T x_t\,dt}{2 + \alpha T}$$

$$\text{var}\,\hat{\beta} = \frac{\sigma^2}{\alpha^2 T + 2\alpha} \longrightarrow \frac{f(0)}{T}$$

Ex. 9. Suppose as in Ex. 8, except that $L(\theta) = \theta^2 + a\theta + b$. Show from the fact that

$$\begin{aligned} K(0, \theta) &= \theta^{-1} L(0)[L(\theta) - e^{-\theta T} L(-\theta)] \\ &= L(0)[(\theta^2 + b)(1 - e^{-\theta T})\theta^{-1} + a(1 + e^{-\theta T})] \end{aligned}$$

that

$$\hat{\beta} = \frac{b\int_0^T x_t\,dt + b(x_T' - x'_0) + a(x_T + x_0)}{bT + 2a}$$

$$\text{var}\,\hat{\beta} = \frac{\sigma^2}{b(bT + 2a)}$$

Ex. 10. Show that for the case of Ex. 8, general $L(\theta)$

$$\text{var}\ \hat{\beta} = \frac{\sigma^2}{L(0)[TL(0) + 2L'(0)]} \longrightarrow \frac{f(0)}{T}$$

Ex. 11. If $x_t = \beta t + \xi_t$, $L(\theta) = \theta + \alpha$, then show from the expression for $\left[\frac{\partial^2 K(\phi, \theta)}{\partial \phi \partial \theta}\right]_{\phi=0}$ that

$$\hat{\beta} = \frac{1}{c}[\alpha^2 \int_0^T t x_t\, dt + (1 + \alpha T) x_T - x_0]$$

$$\text{var}\ \hat{\beta} = \frac{\sigma^2}{c} = \frac{\sigma^2}{(\alpha^2 T^3/3) + \alpha T^2 + T} \longrightarrow \frac{f(0)}{\int_0^T t^2 dt}$$

Ex. 12. Consider the case $x_t = \beta_1 + \beta_2(t - T/2) + \xi_t$, $L(\theta) = \theta + \alpha$.

8.5 Accumulated Processes

One might say that a polynomial of degree $p - 1$ is a solution of the equation

$$\Delta^p x_t = 0 \tag{1}$$

where Δ is the difference operator:

$$\Delta x_t = x_t - x_{t-1} \tag{2}$$

A rather more general and flexible idea than a simple polynomial would be the solution of

$$\Delta^p x_t = \zeta_t \tag{3}$$

where $\{\zeta_t\}$ is a stationary process. A process generated this way we shall term an *accumulated process.* These arise naturally in processes generated by linear mechanisms with little or no damping. In fact, relation (3) is best regarded as a special case of an autoregression with correlated input

$$\sum_0^p a_j x_{t-j} = \zeta_t \tag{4}$$

For stability the zeros of $A(z) = \sum a_j z^j$ must lie outside the unit circle: if these zeros approach the unit circle (in case (3) there is a p-fold zero at $z = 1$), then the process will behave in an evolutive fashion.

The process $\{x_t\}$ generated by (3), starting off from a definite set of initial conditions, will show two types of evolutive behaviour: a polynomial trend, with coefficients determined by the initial values of x, and an indefinite increase in variance.

This type of process, first considered in detail by Yaglom (1955), seems much more natural than the assumption of a simple polynomial superimposed upon a stationary process. It is the model generating the process which is postulated, rather than the final process itself; the model is a plausible limiting case of the conventional linear model (4); and conventional prediction theory can be extended to cover it.

Furthermore, we shall see that these generalised prediction formulae are automatically exact for solutions of (1); i.e. for polynomials of degree up to $p - 1$.

We shall need the following auxiliary result:

Theorem 1

Let $Q(z)$ be a function of z analytic in $\rho < |z| < \rho^{-1}$, and let θ be a number such that $|\theta| < 1$. Then

$$R(z) = (1 - \theta z)^p \left[\frac{Q(z)}{(1 - \theta z)^p} \right]_+ = \Pi_p(z) + [Q(z)]_+ \qquad (5)$$

where $\Pi_p(z)$ is a polynomial in z of degree $p - 1$, so chosen that the differential coefficients of orders $0, 1, 2 \ldots p - 1$ of $R(z)$ are respectively equal to those of $Q(z)$ at $z = \theta^{-1}$. The operation $[\ ]_+$ refers to expansions on the unit circle.

We shall loosely write this last relation between R and Q as

$$R(z) = Q(z) + O[(1 - \theta z)^p] \qquad (6)$$

To prove the theorem, let us decompose Q:

$$Q = [Q]_+ + [Q]_- = Q_+ + Q_- \qquad (7)$$

say, and let the contributions of these two parts to R be denoted by R_+ and R_- respectively. Then $R_+ = Q_+$, so that the theorem is trivial for this component. The other component, Q_-, will be analytic in $|z| > \rho$, and so can be expanded in a partial Taylor series about $z = \theta^{-1}$:

$$Q_-(z) = \sum_0^{p-1} \frac{Q_-^{(j)}(\theta^{-1})}{j!} (z - \theta^{-1})^j + (z - \theta^{-1})^p H(z) \qquad (8)$$

The term $H(z)$ will also have no singularities in $|z| > \rho$, because the only possible such singularity would be at $z = \theta^{-1}$, but if this were present the remainder in (8) would not be $O(z - \theta^{-1})^p$, which in fact we know it to be, (8) being a Taylor expansion. Hence $H+ = 0$, and

$$R_- = \sum_0^{p-1} \frac{Q_-^{(j)}(\theta^{-1})}{j!} (z - \theta^{-1})^j \qquad (9)$$

Hence the theorem is proved, and $\Pi_p(z)$ is identified with R_-, and given explicitly by (9).

Suppose that $\gamma(z)$ is the generating function of the coefficients determining $\hat{x}_{t+\nu}$ in terms of $x_t, x_{t-1} \ldots$ (or $\hat{u}_{t+\nu}$ in terms of $x_s = u_s + \eta_s$ $(s \leqslant t)$, in the case of noise-corrupted signals).

Theorem 2

If, in the sense of equation (6)

$$\gamma(z) = z^{-\nu} + O[(1 - \theta z)^p] \qquad (10)$$

then the predictor is exact for the sequences $x_t = t^j \theta^t$ $(j = 0, 1, \ldots p - 1)$.

By (10) we mean that the jth differentials of $\gamma(z)$ and $z^{-\nu}$ are equal at $z = \theta^{-1}$, $(j = 0, 1 \ldots p-1)$. The same must then be true of $z^{-\nu-t}$ and $\gamma(z)z^{-t}$, so that

$$(t+\nu)_j\,\theta^{t+\nu} = \sum_0^\infty \gamma_s(t-s)_j\theta^{t-s} \quad (j = 0, 1, \ldots p-1) \quad (11)$$

where $t_j = t(t+1)(t+2)\ldots(t+j-1)$. Thus the predictor is exact for sequences $t_j\theta^t$, and so for sequences $t^j\theta^t$.

From these two theorems we obtain the following generalisation of the result of Exs. (3.3.9) and (3.3.10).

Theorem 3

If $g_{uu}(z)$ has $z = \theta^{-1}$ as a p-fold pole ($|\theta| < 1$), but $g_{\eta\eta}(\theta^{-1})$ is finite, then the l.l.s. predictor $\hat{u}_{t+\nu}$ in terms of $x_s = u_s + \eta_s$ ($s \leqslant t$) is exact for sequences $u_t = t^j\theta^t$ $(j = 0, 1 \ldots p-1)$.

Consider first the case of undisturbed prediction, $\eta = 0$, so that $\gamma(z)$ is given by (3.2.7). Then we can write

$$B(z) = \frac{C(z)}{(1-\theta z)^p} \quad (12)$$

where $C(z)^{-1}$ is regular at θ^{-1}. Thus

$$\begin{aligned}\gamma(z) &= \frac{(1-\theta z)^p}{C(z)}\left[\frac{C(z)}{z^\nu(1-\theta z)^p}\right]_+ \\ &= z^{-\nu} + O[(1-\theta z)^p] \end{aligned} \quad (13)$$

by Th. 1. Application of Th. 2 proves our assertion.
For the case $\eta \neq 0$, where there is superimposed "noise", then $\gamma(z)$ is given by (6.1.13), and one further point is involved. We have

$$\begin{aligned}\sigma^2 B(z)B(z^{-1}) &= g_{uu}(t) + g_{\eta\eta}(z) \\ &= \frac{|C(z)|^2}{[(1-\theta z)(1-\theta z^{-1})]^p} + g_{\eta\eta}(z) \\ &= \frac{|D(z)|^2}{[(1-\theta z)(1-\theta z^{-1})]^p} \end{aligned} \quad (14)$$

say. Thus

$$\begin{aligned}\gamma(z) &= \frac{1}{\sigma^2 B(z)}\left[\frac{g_{uu}(z)}{z^\nu B(z^{-1})}\right]_+ \\ &= \frac{(1-\theta z)^p}{D(z)}\left[\frac{|C(z)|^2}{z^\nu D(z^{-1})(1-\theta z)^p}\right]_+ \\ &= \frac{|C|^2}{z^\nu |D|^2} + O[(1-\theta z)^p] \\ &= \frac{|C(z)|^2}{z^\nu[|C(z)|^2 + |1-\theta z|^{2p} g_{\eta\eta}(z)]} + O[(1-\theta z)^p] \\ &= z^{-\nu} + O[(1-\theta z)^p] \end{aligned} \quad (15)$$

so that the result follows again.

Suppose now that the observations x_t are generated in the following fashion: that

$$x_t = u_t + \eta_t \tag{16}$$

where $\{u_t\}$ and $\{\eta_t\}$ are mutually uncorrelated, $g_{\eta\eta}(\theta^{-1})$ is finite, and that

$$(1 - \theta U)^p u_t \equiv \sum_0^p \binom{p}{j}(-\theta)^j u_{t-j} = \zeta_t \tag{17}$$

Here $|\theta| < 1$, and $\{\zeta_t\}$ is a stationary process with s.d.f. $g_{\zeta\zeta}(z)$, analytic in some annulus containing the unit circle and such that $g_{\zeta\zeta}(\theta^{-1}) \neq 0$.

Theorem 4

If x_t, u_t are generated by relations (16) *and* (17) *then the generating function for the l.l.s. predictor $\hat{u}_{t+\nu}$ based on $x_t, x_{t-1} \ldots$ is*

$$\gamma(z) = \frac{1}{D(z)}\left[\sum_0^{p-1} \phi_j(1 - \theta z)^j + \left[\frac{g_{\zeta\zeta}(z)}{z^\nu D(z^{-1})}\right]_+\right] \tag{18}$$

where $D(z)$ is the canonical factor of

$$|D(z)|^2 = g_{\zeta\zeta}(z) + |1 - \theta z|^{2p} g_{\eta\eta}(z) \tag{19}$$

and the coefficients ϕ_j are determined from the Taylor expansion about $z = \theta^{-1}$ of

$$\left[\frac{g_{\zeta\zeta}(z)}{z^\nu D(z^{-1})}\right]_- = \sum_0^{p-1} \phi_j(1 - \theta z)^j + \ldots \tag{20}$$

The predictor is exact for the sequences $u_t = t^j\theta^t$ $(j = 0, 1, \ldots p - 1)$.

The proof of the theorem will be clear from that of Ths. 2 and 3, with the identification $|C|^2 = g_{\zeta\zeta}$. The evaluation of the polynomial follows from (9).

Now, by letting θ tend to 1 from below, we obtain predictors for the accumulated process $u_t = \Delta^{-p}\zeta_t$. Despite the fact that the variance of u_t (and so of x_t) increases indefinitely with t in this case, the prediction m.s.e. σ_ν^2 over a finite interval will always be finite. The m.s.e. will not tend to a constant as ν increases however, but behaves as ν^{2p-1} in general.

Ex. 1. If ζ_t in equation (3) has the m.a. representation

$$\zeta_t = \sum_0^\infty b_j \varepsilon_{t-j}$$

show that

$$x_t = \sum b_j^* \varepsilon_{t-j} + \sum_0^{p-1} c_j t^j$$

where the c_j are arbitrary constants, and

$$b_j^* = \sum_{k=0}^{j} \frac{(p + k - 1)!}{(p - 1)!\, k!} b_{j-k}$$

(We suppose that the process has started at some definite time, so that $\varepsilon = 0$ if one goes sufficiently far back in time, and the sum $\Sigma b_j^* \varepsilon_{t-j}$ does not diverge for finite t.) Show that the prediction m.s.e. is

$$\sigma_\nu^2 = \sigma^2(\varepsilon) \sum_0^{\nu-1} (b_j^*)^2 \sim \frac{\sigma^2(\sum_0^\infty b_j)^2 \nu^{2p-1}}{(2p-1)[(p-1)!]^2}$$

if $\sum_0^\infty b_j \neq 0$.

Ex. 2. Show both from Th. 4 and directly from Ex. (3.3.3) that if $\Delta x_t = \zeta_t$, $\zeta_t = \rho\zeta_{t-1} + \varepsilon_t$, where $\{\varepsilon_t\}$ is an uncorrelated sequence, then

$$\hat{x}_{t+\nu} = \rho^\nu x_t + \frac{1-\rho^\nu}{1-\rho}(x_t - \rho x_{t-1})$$

$$\sigma_\nu^2 = \frac{\sigma^2}{(1-\rho)^2}\left[\nu - 2\rho\frac{1-\rho^\nu}{1-\rho} + \frac{\rho^2(1-\rho^{2\nu})}{1-\rho^2}\right]$$

Ex. 3. Suppose that $\Delta u_t = \zeta_t$, $\zeta_t = \rho\zeta_{t-1} + \varepsilon_t$, $x_t = u_t + \eta_t$, where $\{\varepsilon_t\}$, $\{\eta_t\}$ are uncorrelated sequences with respective variances σ^2, $\lambda\sigma^2$. Show that $\hat{u}_{t+\nu}$ is determined by

$$\gamma(z) = \frac{A + Bz}{(1-\beta_1 z)(1-\beta_2 z)}$$

where β_1 and β_2 are the two roots inside the unit circle of

$$1 + \lambda|(1-z)(1-\rho z)|^2 = 0$$

and A and B are determined so that $\gamma(\theta^{-1}) = \theta^\nu$; $\theta = 1, \rho$.

Ex. 4. Generalise Th. 4 to the case (4), when the zeros of $A(z)$ are distributed arbitrarily in $|z| \geqslant 1$.

The analogues of Ths. 1–4 for continuous time are fairly clear: for convenience we give the analogue of Th. 4, but specialised to the case of an accumulated process, corresponding to $\theta = 1$ in Th. 4.

Theorem 4A

Suppose we observe $x_t = u_t + \eta_t$, where

$$D^p u_t = \zeta_t$$

and the processes $\{\zeta_t\}$, $\{\eta_t\}$ are mutually uncorrelated, with s.d.f.'s analytic in a strip containing the real ω axis, and $f_{\zeta\zeta}(0) \neq 0$. Then $\hat{u}_{t+\nu}$ is determined by

$$\gamma(\omega) = \frac{1}{D(\omega)}\left[\pi(\omega) + \left[\frac{e^{i\nu\omega} f_{\zeta\zeta}(\omega)}{D(-\omega)}\right]_+\right]$$

where $D(\omega)$ is the canonical factor (assumed to exist) of

$$D(\omega)D(-\omega) = f_{\zeta\zeta}(\omega) + \omega^{2p} f_{\eta\eta}(\omega)$$

and the polynomial $\pi(\omega)$ is equal to the sum of the first p terms of the Taylor expansion of $[e^{i\nu\omega} f_{\zeta\zeta}/\bar{D}]_-$ about $\omega = 0$.

It is also true that $\gamma(\omega) = e^{i\nu\omega} + O(\omega^p)$, so that the predictor is exact for $u_t = t^j$ $(j = 0, 1, 2 \ldots p-1)$.

Ex. 5. Work the continuous analogue of Ex. 3, and find the condition that the quantities corresponding to β_1 and β_2 be real or complex.

8.6. Exponentially Weighted Moving Averages

As we have remarked before, the notation $\hat{x}_{t+\nu}$ is an ambiguous one: and it is sometimes necessary to modify the expression to $\hat{x}_{t+\nu,\,\nu}$, the predictor of $x_{t+\nu}$ based upon $x_t, x_{t-1}, \ldots$

It has been suggested by various authors (Holt (see Holt *et al.*, 1960); Cox, 1961) that useful predictors could be constructed from the recursive relations of the type

$$\hat{x}_{t+\nu,\,\nu} = \lambda x_t + \mu \hat{x}_{t+\nu-1,\,\nu} \tag{1}$$

where the coefficients λ and μ are chosen in some semi-empiric fashion. The solution of (1) for the prediction in terms of observations is

$$\hat{x}_{t+\nu,\,\nu} = \lambda \sum_0^\infty \mu^j x_{t-j} \tag{2}$$

and it is for this reason that the method is given the title above. However, as we saw in Ex. (3.3.12), these predictors are just what would be obtained on the Wiener theory for a process with rational spectral density,

$$g(z) = \sigma^2 \left|\frac{Q(z)}{P(z)}\right|^2 \tag{3}$$

and the only novelty in the proposal seems to be the computational convenience of (1) over (2).

Thus, if $P(z) = 1 - \alpha z$, $Q(z) = 1 - \beta z$, then the Wiener predictor obeys the relation

$$\hat{x}_{t+\nu,\,\nu} = (\alpha - \beta)\alpha^\nu x_t + \beta \hat{x}_{t+\nu-1,\,\nu} \tag{4}$$

which is just of the form (1). A more general example of such recursions is given in Ex. (3.3.12).

Sometimes the coefficients λ and μ are chosen so that the prediction is exact for $x_t =$ const. (or for x a polynomial of given degree in more general cases). This, we know from section (5), will be automatically achieved in the conventional l.l.s. theory if $P(z)$ has zeros at $z = 1$. For instance, if $\alpha = 1$ in (4) above, then

$$\hat{x}_{t+\nu,\,\nu} = (1 - \beta)x_t + \beta \hat{x}_{t+\nu-1,\,\nu} \tag{5}$$

which is exact for $x_t =$ const.

Ex. 1. Show that if $P(z) = (1 - \alpha_1 z)(1 - \alpha_2 z)$, $Q(z) = 1 - \beta z$, then the l.l.s. predictor obeys a relation

$$\hat{x}_{t+\nu,\,\nu} - \beta \hat{x}_{t+\nu-1,\,\nu} = \lambda x_t + \mu x_{t-1}$$

where λ and μ are such that the predictor is exact for sequences α_1^t, α_2^t (cf. Ex. (3.3.11)).

CHAPTER 9

MULTIVARIATE PROCESSES

It is probably not unfair to say that 90% of the literature is devoted to univariate rather than to multivariate processes, while in genuine applications the frequencies are reversed. One almost always works with systems whose states must be described by several variables, and ultimately one must be able to cope with the multivariate situation, although it is reasonable that ideas and methods should be developed first for the univariate case.

In fact, with the appropriate formalism, almost all our previous work generalises immediately. There is one exception, however; the canonical factorisation of the s.d.f. $g(z)$, or $f(\omega)$, which yields the canonical m.a. representation of the process. The s.d.f. now becomes a matrix of s.d.f.'s, and its factorisation presents a new problem. Prerequisites: Chs. 2–7.

9.1 Projection on the Infinite Sample

Suppose that the process we observe is $\{\mathbf{x}_t\}$, where $\mathbf{x}_t$ is a column vector of component random variates: $(x_{1t}, x_{2t}, \ldots x_{rt})$. We shall assume the process stationary and purely non-deterministic, with a spectral density matrix (s.d.m.) defined by equation (2.1.15) and that the elements of both $\mathbf{g}$ and $\mathbf{g}^{-1}$ are analytic in some annulus $\rho < |z| < \rho^{-1}$ (this is for the discrete time case). These assumptions are quite unnecessarily strong for the present section, as one can see by comparison with Ch. 5, but they are the convenient ones for the next section.

Suppose that one wished to predict a column vector of variates $\mathbf{y} = (y_1, y_2, \ldots y_m)$. As before, this may be a member of a stationary sequence, in which case we simply append the subscript t to each variate. In either case we define the *cross-spectral density matrix* (also to be abbreviated to s.d.m.)

$$\mathbf{g}_{yx}(z) = \mathbf{f}_{yx}(\omega) = [\sum_s \operatorname{cov}(y_j, x_{k,t-s})z^s] \tag{1}$$

Let the l.l.s. predictor be written

$$\hat{y}_j = \sum_k \sum_u \gamma_{jku} x_{k,t-u} \tag{2}$$

or

$$\hat{\mathbf{y}} = \sum \gamma_u \mathbf{x}_{t-u} \tag{3}$$

where $\boldsymbol{\gamma}_u$ is the $m \times r$ matrix with elements γ_{jku}. Then by demanding that $\hat{\mathbf{y}} - \mathbf{y}$ be uncorrelated with $\mathbf{x}_{t-s}$ (Ex. 4.1.3) we find that

$$E(\mathbf{y}\mathbf{x}'_{t-s}) = \sum_u \boldsymbol{\gamma}_u\, E(\mathbf{x}_{t-u}\mathbf{x}'_{t-s}) \tag{4}$$

Multiplying by z^s and adding we thus obtain

$$\mathbf{g}_{\mathbf{yx}}(z) = \boldsymbol{\gamma}(z)\mathbf{g}_{\mathbf{xx}}(z) \tag{5}$$

where

$$\boldsymbol{\gamma}(z) = \sum \boldsymbol{\gamma}_j z^j \tag{6}$$

Solution (5.1.3) thus now becomes

$$\boldsymbol{\gamma}(z) = \mathbf{g}_{\mathbf{yx}}(z)\mathbf{g}_{\mathbf{xx}}(z)^{-1} \tag{7}$$

For the more general case considered in section (5.1) one would expect that it could be written

$$\boldsymbol{\gamma}(z) = [d\mathbf{F}_{\mathbf{yx}}(\omega)][d\mathbf{F}_{\mathbf{xx}}(\omega)]^{-1} \tag{8}$$

with a similar relation in the continuous time case for

$$\boldsymbol{\gamma}(\omega) = \int \boldsymbol{\gamma}_s e^{-iws}\, ds \tag{9}$$

where

$$\hat{\mathbf{y}} = \int \boldsymbol{\gamma}_s \mathbf{x}_{t-s}\, ds \tag{10}$$

If $\mathbf{y}$ really is an element, $\mathbf{y}_t$, of a process which is jointly stationary with $\{\mathbf{x}_t\}$, then the s.d.m. of the deviations $\boldsymbol{\delta}_t = \hat{\mathbf{y}}_t - \mathbf{y}_t$ will be given by

$$\mathbf{f}_{\delta\delta} = \mathbf{f}_{\mathbf{yy}} - \mathbf{f}_{\mathbf{yx}}\mathbf{f}_{\mathbf{xx}}^{-1}\mathbf{f}_{\mathbf{xy}} \tag{11}$$

If $\mathbf{y}$ is a single variable then one can only say that

$$E(\boldsymbol{\delta}\boldsymbol{\delta}') = E(\mathbf{y}\mathbf{y}') - \frac{1}{2\pi}\int \mathbf{f}_{\mathbf{yx}}\mathbf{f}_{\mathbf{xx}}^{-1}\mathbf{f}_{\mathbf{xy}}\, d\omega \tag{12}$$

9.2 Projection on the Semi-infinite Sample

In order to treat the case where $\hat{\mathbf{y}}$ is based upon $(\mathbf{x}_s;\, s \leqslant t)$, we have to modify (1.7) in the same way that the solution (5.1.3) was modified to become (6.1.3).

Suppose that we can achieve a canonical factorisation of the s.d.m.

$$\mathbf{g}_{\mathbf{xx}}(z) = \mathbf{P}(z)\mathbf{P}(z)^{\dagger} \tag{1}$$

where $\mathbf{P}(z)$ is not identically singular, and where the Laurent expansions of $\mathbf{P}(z)$ and $\mathbf{P}(z)^{-1}$ on $|z| = 1$ (i.e. expansions element by element) contain no negative powers of z. By $\mathbf{P}(z)^{\dagger}$ we mean the transpose of $\mathbf{P}(z^{-1})$.

A slight modification of (1) is

$$\mathbf{g}_{\mathbf{xx}}(z) = \mathbf{B}(z)\mathbf{V}\mathbf{B}(z)^{\dagger} \tag{2}$$

where $\mathbf{V}$ is independent of z, and is chosen so as to make $\mathbf{B}_0 = \mathbf{I}$, where $\mathbf{B}_0$ is the absolute coefficient in the expansion

$$\mathbf{B}(z) = \sum_0^\infty \mathbf{B}_j z^j \tag{3}$$

Thus $\mathbf{V} = \mathbf{P}_0\mathbf{P}_0'$. Representations (1) and (2) correspond to closely-related moving average representations of the process:

$$\mathbf{x}_t = \sum_0^\infty \mathbf{P}_j \boldsymbol{\eta}_{t-j} = \sum_0^\infty \mathbf{B}_j \boldsymbol{\varepsilon}_{t-j} \tag{4}$$

where $\boldsymbol{\varepsilon}_t = \mathbf{P}_0\boldsymbol{\eta}_t$. Here

$$E[\boldsymbol{\eta}_s\boldsymbol{\eta}_t'] = \delta_{st}\mathbf{I} \tag{5}$$

$$E[\boldsymbol{\varepsilon}_s\boldsymbol{\varepsilon}_t'] = \delta_{st}\mathbf{P}_0\mathbf{P}_0' = \delta_{st}\mathbf{V} \tag{6}$$

so the form which one uses depends upon whether one chooses the normalisations of standardised "residuals", (5), or of $\mathbf{B}_0 = \mathbf{I}$.

If one assumes the existence of representation (1), then it follows by the arguments of either of sections (3.7) or (6.1), that

$$\boldsymbol{\gamma}(z) = [\mathbf{g}_{yx}(z)(\mathbf{P}(z)^\dagger)^{-1}]_+\mathbf{P}(z)^{-1} \tag{7}$$

Alternatively,

$$\boldsymbol{\gamma} = [\mathbf{g}_{yx}(\mathbf{B}^\dagger)^{-1}\mathbf{V}^{-1}]_+\mathbf{B}^{-1} = [\mathbf{g}_{yx}(\mathbf{B}^\dagger)^{-1}]_+\mathbf{V}^{-1}\mathbf{B}^{-1} \tag{8}$$

Ex. 1. Show for the case of pure prediction, when $\mathbf{y}_t = \mathbf{x}_{t+\nu}$, that

$$\boldsymbol{\gamma}(z) = [z^{-\nu}\mathbf{B}(z)]_+\mathbf{B}(z)^{-1}$$

and that, if $\boldsymbol{\delta} = \hat{\mathbf{x}}_{t+\nu} - \mathbf{x}_{t+\nu}$

$$E(\boldsymbol{\delta\delta}') = \sum_0^{\nu-1} \mathbf{B}_j\mathbf{V}\mathbf{B}'_j = \sum_0^{\nu-1} \mathbf{P}_j\mathbf{P}_j'$$

The real problem is the achievement of the representation (1) or (2). In the univariate case, if $\log g(z)$ had a Laurent expansion $\sum c_j z^j$ on $|z| = 1$, then one could say immediately that

$$B(z) = e^{\sum_1^\infty c_j z^j} \tag{9}$$

was the desired factor. This will no longer be true in so obvious a sense, because it will no longer be true in general that

$$e^{\sum_{-\infty}^{\infty} c_j z^j} = e^{\sum_1^\infty c_j z^j} e^{c_0} e^{\sum_{-\infty}^{-1} c_j z^j} \tag{10}$$

since the c_j are now matrices which will not in general commute with one another. In fact, despite the attention of several distinguished workers (among others, Wiener and Masani (1957, 1958); Helson and Lowdenslager (1958)), the determination of the canonical factorisation is still an unsolved problem.

For the case of rational spectral densities we did not need to resort to (9), but could simply factorise numerator and denominator from a knowledge of their zeros. Again, this method does not generalise to the multivariate case in a simple fashion, but it does generalise (see section (3)).

The order of the factors in (1) was dictated by the fact that we were predicting from the past: had we based $\hat{\mathbf{y}}$ upon a "future" sample $(\mathbf{x}_s; s \geqslant t)$, we should have had to find a representation

$$\mathbf{g}_{\mathbf{xx}}(z) = \mathbf{P}_f(z)^\dagger \mathbf{P}_f(z) = \mathbf{B}_f(z)^\dagger \mathbf{V}_f \mathbf{B}_f(z) \tag{11}$$

where $\mathbf{P}_f$ and $\mathbf{B}_f$ have the same defining properties as $\mathbf{P}$ and $\mathbf{B}$. It is a characteristic feature of the multivariate case that, in general, $\mathbf{P}_f \neq \mathbf{P}$.

9.3 Canonical Factorisation of a Rational Spectral Density Matrix

We shall consider consecutively the cases where the process is known to be generated by a pure autoregression, a pure moving average (of finite order, in both cases), and by a process of mixed type.

We shall denote the s.d.m. simply by $\mathbf{g}(z)$. If the process is a pure a.r., order p, then there must be a representation

$$\mathbf{g}(z) = \mathbf{A}(z)^{-1}\mathbf{V}(\mathbf{A}(z)^\dagger)^{-1} \tag{1}$$

where $\mathbf{A}(z)$ is a polynomial

$$\mathbf{A}(z) = \sum_0^p \mathbf{A}_j z^j \tag{2}$$

for which $\mathbf{A}_0 = \mathbf{I}$, and all zeros of $|\mathbf{A}(z)|$ lie outside the unit circle. The problem is, to find $\mathbf{A}$, given $\mathbf{g}$. We see from (1) that the function

$$\mathbf{A}(z)\mathbf{g}(z) = \mathbf{V}(\mathbf{A}(z)^\dagger)^{-1} \tag{3}$$

contains no positive powers of z in its Laurent expansion on $|z| = 1$. Thus

$$\sum_0^p \mathbf{A}_k \mathbf{\Gamma}_{j-k} = 0 \qquad (j = 1, 2, \ldots) \tag{4}$$

These are just the Yule–Walker relations for the multivariate a.r. Provided at least that there are no degeneracies in the system, we can use equations (4) for $j = 1, 2, \ldots p$ to determine $\mathbf{A}_1, \mathbf{A}_2, \ldots \mathbf{A}_p$. The factor $\mathbf{A}(z)$ is thus determined, and one can then evidently solve for $\mathbf{V}$ from equation (1).

Ex. 1. Suppose that $\mathbf{g}(z)$ has a known representation of form (1):

$$\mathbf{g} = (\mathbf{I} - z\boldsymbol{\alpha})^{-1}\mathbf{V}(\mathbf{I} - z^{-1}\boldsymbol{\alpha}')^{-1}$$

where $\boldsymbol{\alpha}$ is a matrix with all its eigenvalues inside the unit circle. Show that there is a corresponding representation of type (2.11):

$$\mathbf{g} = (\mathbf{I} - z^{-1}\boldsymbol{\alpha}_f')^{-1}\mathbf{V}_f(\mathbf{I} - z\boldsymbol{\alpha}_f)^{-1}$$

where

$$\boldsymbol{\alpha}_f = \mathbf{R}^{-1}\boldsymbol{\alpha}\mathbf{R}$$
$$\mathbf{V}_f = \mathbf{R} - \mathbf{R}\boldsymbol{\alpha}'\mathbf{R}^{-1}\boldsymbol{\alpha}\mathbf{R}$$
$$\mathbf{R} = \sum_0^\infty \boldsymbol{\alpha}^j\mathbf{V}(\boldsymbol{\alpha}')^j$$

Even if $\mathbf{g}(z)$ is not exactly of the form (1), one can still use this technique of autoregressive fitting to obtain an approximate factorisation of the s.d.m., as in section (3.4). It can be shown that the fitted $\mathbf{A}(z)$ will always have the property we wish of it; that all zeros of $|\mathbf{A}(z)|$ lie in $|z| > 1$ (provided only that there are no exact relations among the variates $\mathbf{x}_t, \mathbf{x}_{t-1}, \ldots \mathbf{x}_{t-p}$ included in the fitted a.r.; otherwise there may be roots on $|z| = 1$).

Ex. 2. Show that the recursive method of fitting described in equations (3.4.7)–(3.4.12) generalises to the present case as follows. If one considers simultaneously the fitting of "backward" and "forward" a.r.s of order p:

$$\sum_{k=0}^{p} \mathbf{A}_k^{(p)}\mathbf{x}_{t-k} = \boldsymbol{\varepsilon}_t^{(p)}$$
$$\sum_{k=0}^{p} \mathbf{A}_{kf}^{(p)}\mathbf{x}_{t+k} = \boldsymbol{\varepsilon}_{tf}^{(p)}$$

where $\mathbf{A}_0^{(p)} = \mathbf{A}_{0f}^{(p)} = \mathbf{I}$, and

$$\mathbf{V}_p = E[\boldsymbol{\varepsilon}_t^{(p)}\boldsymbol{\varepsilon}_t^{(p)\prime}] = \sum_0^p \mathbf{A}_k^{(p)}\boldsymbol{\Gamma}_{-k}$$
$$\mathbf{V}_{pf} = E[\boldsymbol{\varepsilon}_{tf}^{(p)}\boldsymbol{\varepsilon}_{tf}^{(p)\prime}] = \sum_0^p \mathbf{A}_{kf}^{(p)}\boldsymbol{\Gamma}_k$$
$$\boldsymbol{\Delta}_p = \sum_0^p \mathbf{A}_k^{(p)}\boldsymbol{\Gamma}_{p+1-k}$$
$$\boldsymbol{\Delta}_{pf} = \sum_0^p \mathbf{A}_{kf}^{(p)}\boldsymbol{\Gamma}_{k-p-1}$$

then

$$\mathbf{A}_k^{(p+1)} = \mathbf{A}_k^{(p)} + \mathbf{A}_{p+1}^{(p+1)}\mathbf{A}_{p+1-k,\,f}^{(p)}$$
$$\mathbf{A}_{kf}^{(p+1)} = \mathbf{A}_{kf}^{(p)} + \mathbf{A}_{p+1,\,f}^{(p+1)}\mathbf{A}_{p+1-k}^{(p)} \qquad (k = 1, 2, \ldots p)$$
$$\mathbf{A}_{p+1}^{(p+1)} = -\,\boldsymbol{\Delta}_p\mathbf{V}_{pf}^{-1}$$
$$\mathbf{A}_{p+1,\,f}^{(p+1)} = -\,\boldsymbol{\Delta}_{pf}\mathbf{V}_p^{-1}$$

It is interesting that one must fit the two types of a.r. if one is to have a recursion; i.e. one must determine the two types of factorisation simultaneously.

Consider now the m.a. case: we assume that $\mathbf{g}(z)$ has a finite Laurent expansion, containing the powers $z^{-q} \ldots z^q$, and we wish to determine the factor $\mathbf{B}(z)$ in (2.2), which will now be a polynomial in z of degree q. Formally, the technique is simple, one forms the matrix $\mathbf{g}(z)^{-1}$ and then applies the method used for the a.r. case. Thus, if one knows the elements of $\mathbf{g}(z)$ explicitly as functions of z, and if $|\mathbf{g}(z)|$ has no zeros on the unit circle, then one can determine a Laurent expansion on $|z| = 1$;

$$\mathbf{g}(z)^{-1} = \sum \mathbf{g}^{(j)}(z) \tag{5}$$

and, corresponding to equations (4), one has

$$\sum_{k=0}^{q} \mathbf{g}^{(j-k)}\mathbf{B}_k = 0 \qquad (j = 1, 2, \ldots) \tag{6}$$

If there are no degeneracies then the equations (6) for $j = 1, 2, \ldots q$ will determine $\mathbf{B}_1, \mathbf{B}_2, \ldots \mathbf{B}_q$.

Consider now the mixed case: suppose we know that $\mathbf{g}(z)$ can be represented

$$\mathbf{g}(z) = \mathbf{A}(z)^{-1}\mathbf{B}(z)\mathbf{V}\mathbf{B}(z)^{\dagger}(\mathbf{A}(z)^{\dagger})^{-1} \tag{7}$$

where $\mathbf{A}$ and $\mathbf{B}$ are polynomials in z of degrees p and q respectively, with all roots in $|z| > 1$. This corresponds to the existence of a relation

$$\sum_{0}^{p} \mathbf{A}_s \mathbf{x}_{t-s} = \sum_{0}^{q} \mathbf{B}_s \boldsymbol{\varepsilon}_{t-s} \tag{8}$$

where $\boldsymbol{\varepsilon}_t$ obeys (2.6).

Relations (4) will still hold for $j > q$, and, in the absence of degeneracies, we can use the relations for $j = q + 1, q + 2, \ldots p + q$ to determine $\mathbf{A}(z)$. One can then calculate $\mathbf{BVB}^{\dagger}$, and determine $\mathbf{B}$ from the inverse of this matrix, as in (6).

There have been several gaps in our argument: we have not shown that any s.d.m. with rational elements can be expressed in the form (7); we have not considered the effect of degeneracies in the equations systems (4) or (6); nor have we demonstrated that the fitting of multivariate a.r.'s by equations (4) will give an approximate factorisation which is nearly of the kind desired. These are all topics which demand a more extended treatment than is appropriate here; the material we have given should, however, be sufficient for a first attack upon most practical problems.

9.4 Projection on the Finite Sample

We shall very briefly describe the analogues of results (7.1.10), (7.1.11), and (7.2.1), omitting proofs, so that this section is virtually an exercise for the reader. The generalisation of these results is not entirely trivial, since one has to call upon both the canonical factorisations, (2.1) and (2.11). This is not surprising, when one considers that a finite sample is truncated both to the future and to the past.

Suppose that we are in the a.r. case:

$$\mathbf{V} = \mathbf{A}^{-1}\, \mathbf{V}(\mathbf{A}^{\dagger})^{-1} = (\mathbf{A}_f^{\dagger})^{-1}\mathbf{V}_f\mathbf{A}_f^{-1} \tag{1}$$

where both $\mathbf{A}$ and $\mathbf{A}_f$ are of degree p. (It is in fact true that if one type of representation holds, then so does the other, both of the same degree, p.) Suppose that one is considering an equation system

$$\sum_{k=0}^{n-1} \boldsymbol{\lambda}_k \boldsymbol{\Gamma}_{j-k} = \boldsymbol{\mu}_j \qquad (j = 0, 1, \ldots n-1) \tag{2}$$

where λ_j, μ_j are r-columned matrices. Then the analogue of (7.1.10) is

$$\boldsymbol{\lambda}(z) = \boldsymbol{\lambda} = [\boldsymbol{\mu}\mathbf{A}^{\dagger}]_{+}\mathbf{V}^{-1}\mathbf{A} - [\boldsymbol{\mu}\mathbf{A}_f]_n \mathbf{V}_f^{-1}\mathbf{A}_f^{\dagger} \tag{3}$$

That of (7.1.11) is

$$\mathbf{K}(z,w) = (1 - zw)^{-1}[\mathbf{A}(w)'\mathbf{V}^{-1}\mathbf{A}(z) - (zw)^n \mathbf{A}_f(w^{-1})\mathbf{V}_f^{-1}\mathbf{A}_f(z^{-1})'] \tag{4}$$

$\mathbf{K}(z, w)$ is itself an $r \times r$ matrix, with the property

$$K(w, z) = K(z, w)' \tag{5}$$

For the m.a. case we shall have

$$\mathbf{g}(z) = \mathbf{P}\mathbf{P}^{\dagger} = \mathbf{P}_f^{\dagger}\mathbf{P}_f \tag{6}$$

where $\mathbf{P}$ and $\mathbf{P}_f$ are polynomials in z of degree q, say. For the equation system (2) the analogue of (7.2.1) is

$$\begin{aligned} \boldsymbol{\lambda}(z) &= [\boldsymbol{\mu}(z) + \sum_{-q}^{-1} \boldsymbol{\mu}_j z^j + \sum_{n}^{n+q-1} \boldsymbol{\mu}_j z^j]\mathbf{g}(z)^{-1} \\ &= \mathbf{M}(z)\mathbf{g}(z)^{-1} \end{aligned} \tag{7}$$

say. The unknown second and third terms in $\mathbf{M}(z)$ are again determined from the fact that $\boldsymbol{\lambda}(z)$ is polynomial. Suppose the zeros z_ν of $|\mathbf{g}(z)|$ are simple, and that

$$\lim_{z \to z_\nu} (z - z_\nu)\mathbf{g}(z)^{-1} = \boldsymbol{\xi}_\nu \boldsymbol{\eta}_\nu^{\dagger} \tag{8}$$

where $\boldsymbol{\xi}$ and $\boldsymbol{\eta}$ are r-vectors. Then we must have

$$\mathbf{M}(z_\nu)\boldsymbol{\xi}_\nu = 0 \qquad (\nu = 1, 2, \ldots 2rq) \tag{9}$$

These relations are just sufficient to determine the $2q$ unknown $\boldsymbol{\mu}_j$ in (7).

It is interesting that we have not needed to use the explicit factorisation (6). However, the roots z_ν can be arranged in pairs of reciprocals, the ones outside the unit circle being the zeros of $|\mathbf{P}|$ and of $|\mathbf{P}_f|$. If the equation system (2) were extended to infinity, either in the direction of increasing or of decreasing j, then the distinction between the two groups of roots would become important, and one would use the dissection (6).

9.5 Continuous Time

To ring all the changes of discrete or continuous time, sample size, deterministic components, and evolutive behaviour anew for the multivariate case would tire both reader and author. In this section we shall describe the most useful piece of work, the formal factorisation of a rational s.d.m., again leaving the details as an exercise to the reader. For a thorough treatment see Yaglom (1960).

Suppose that the s.d.m. $\mathbf{f}(\omega)$ is known as a matrix function of ω, and known to be representable in the form

$$\begin{aligned}\mathbf{f}(\omega) &= \mathbf{L}(i\omega)^{-1}\mathbf{M}(i\omega)\mathbf{M}(-i\omega)'(\mathbf{L}(-i\omega)')^{-1} \\ &= \mathbf{L}^{-1}\mathbf{M}\mathbf{M}^{\dagger}(\mathbf{L}^{\dagger})^{-1} \qquad (1)\end{aligned}$$

where $\mathbf{L}(\zeta) = \sum_0^l \mathbf{L}_j\zeta^j$ and $\mathbf{M}(\zeta)$ are matrix polynomials whose determinants have zeros only in the left half-plane (i.e. with negative real part). The problem is to determine $\mathbf{L}$ and $\mathbf{M}$.

Corresponding to (3.4), one has

$$\mathbf{L}\left(\frac{d}{ds}\right)\mathbf{\Gamma}_s = 0 \qquad (s > 0) \qquad (2)$$

this being true even if $\mathbf{M}$ is of higher degree than the zeroth. One thus has, by differentiation of (2)

$$\sum_k \mathbf{L}_k\left(\frac{d}{ds}\right)^{j+k}\mathbf{\Gamma}_s = 0 \qquad \left(\begin{matrix} s > 0, \\ j = 0, 1, 2, \ldots \end{matrix}\right) \qquad (3)$$

(Note that if (1) is true then $\mathbf{\Gamma}_s$ is differentiable infinitely often for $s \neq 0$.)

One can fix on a value s ($0+$, say) and use the resulting equation system (3) to determine the $\mathbf{L}_j$. This may not be as simple as for the discrete case, because it is not necessarily true that either $\mathbf{L}_0$ or $\mathbf{L}_l$ is of full rank, and so can be normalised to $\mathbf{I}$. Supposing $\mathbf{L}$ determined, one can then determine $\mathbf{M}\mathbf{M}^{\dagger}$, and thence determine the Fourier representation of its inverse:

$$(\mathbf{M}\mathbf{M}^{\dagger})^{-1} = \int e^{-i\omega s}\mathbf{H}(s)\, ds \qquad (4)$$

From this it follows, as in (2), that

$$\sum_k \left(\frac{d}{ds}\right)^k \mathbf{H}(s)\mathbf{M}_k = 0 \qquad (s > 0)$$

which again yields a set of equations for the $\mathbf{M}_k$, as in (3).

CHAPTER 10
REGULATION

As we emphasised in the first chapter, prediction is rarely an end in itself. In most cases the predicted value, once obtained, is used to initiate or modify a course of action. For example, the predicted course of a target plane is immediately used to help aim anti-aircraft equipment; sales forecasts are used to help plan production.

In this larger context the problem of prediction appears only as incidental, and the central problem is that of *regulation*, i.e. of using past values to determine present action in such a way that the future course of the process is as near as possible to the desired one.

Classical servomechanism theory is concerned with just such situations, the specific applications usually being the stabilisation of mechanical or electrical devices, or of continuous chemical processes (see, for instance, James, Nichols and Phillips (1947), Truxal (1955), Tsien (1954)). This theory originally contained few statistical ideas, but, particularly since the development of radar tracking and fire control in the Second World War, it has been found necessary to consider systems with statistical inputs. In fact, the trend has since continued to the point where statistical criteria of performance (such as mean square error) are being used for the optimisation of design, although they have by no means replaced the older methods, which sought a semi-subjective compromise between demands on stability, frequency and transient responses, dynamic lags, etc, It is sometimes said that statistical design takes too little account of local stability; we discuss this point in section (9).

Another field where the synthesis of statistical and control ideas may prove fruitful is that of economic regulation (Tustin (1953), Holt *et al.* (1960)). Here regulation is achieved, not automatically by a physical device such as a filter and a feedback line, but by a periodic resetting of variables such as production rates, tax rates or level of spending, the new values being determined from observations on such quantities as orders, reserves or level of unemployment.

The most general formulation hitherto of a statistical theory of optimum regulation is R. Bellman's concept of *dynamic programming* (Bellman (1957), (1961)). This work is obviously fundamental, and has already had its successes, but such a general approach will not quickly yield solutions to specific problems.

The method we shall describe in this chapter is a less ambitious one, being restricted to the consideration of linear rules, quadratic criteria of performance (such as mean square deviations) and, in the main, to stationary processes. In fact, the theory amounts to an extension of l.l.s. estimation theory; more particularly, of the Wiener theory, since the Wiener–Hopf technique is used constantly. Like the Wiener theory, it defers the question of the actual physical realisation of a filter: rather than optimising a given filter (or set of relations between variates) with respect to a few parameters, one optimises the whole filter freely, subject only to realisability conditions. As is usually the case, the less the optimal solution is restricted, the easier it is to find.

The first to treat least-square regulation along these lines seems to have been G. C. Newton (1952, 1957). Very similar techniques have been developed later by Holt and other authors (1960), for the construction of linear decision rules in economics. Holt and his co-authors seem to be unaware of Newton's work and their methods show an interesting contrast: instead of transforms, stochastic averages and the Wiener–Hopf technique they use matrices, time averages and the principle of "certainty equivalence". We shall describe both techniques. We do not consider time-varying systems: for an account of statistical methods for such systems see Peterson (1961).

Prerequisites: Chs. 2, 4, 5, 6; also section (8.5) for the sections on non-stationary inputs.

10.1 Notation for Operators

We have consistently used the notation

$$\alpha(z) = \sum \alpha_j z^j \tag{1}$$

in discrete time, and, when talking of generating functions, have used α, $\bar{\alpha}$ and $|\alpha|^2$ to denote $\alpha(z)$, $\alpha(z^{-1})$ and $\alpha(z)\alpha(z^{-1})$, with analogous notations in the vector case. It is convenient now to extend this, so that, if we are talking of the time domain, then α, $\bar{\alpha}$ and $|\alpha|^2$ will be understood to mean $\alpha(U)$, $\alpha(U^{-1})$ and $\alpha(U)\alpha(U^{-1})$, where U is the backward translation operator having the property

$$Ux_t = x_{t-1} \tag{2}$$

Thus, αx_t has the connation

$$\alpha x_t = \alpha(U)x_t = \sum \alpha_j x_{t-j} \tag{3}$$

while $\alpha\bar{\beta}x_t$ would be

$$\alpha\bar{\beta}x_t = \alpha(U)\beta(U^{-1})x_t = \sum_j\sum_k \alpha_j\beta_k x_{t-j+k} \tag{4}$$

the expansions always being those valid on the unit circle.

Analogously, if α_s is a function of continuous time, and

$$\alpha(\omega) = \int e^{-i\omega s}\alpha_s\, ds \tag{5}$$

its Fourier transform, then, if we are talking of the time-domain, α

and $\bar{\alpha}$ will denote the operators $\alpha\left(\frac{d}{idt}\right), \alpha\left(-\frac{d}{idt}\right)$. For example, suppose

$$\alpha(\omega) = \frac{1}{i\omega + \mu} \tag{6}$$

where $Re(\mu) > 0$, so that the Fourier representation of $\alpha(\omega)$ for real ω is

$$\alpha(\omega) = \int_0^\infty e^{-i\omega s - \mu s}\, ds \tag{7}$$

Then by αx_t we mean

$$\alpha x_t = \int_0^\infty e^{-(\mu+D)s} x_t\, ds = \int_0^\infty e^{-\mu s} x_{t-s}\, ds \tag{8}$$

Another convenient notation is the use of (y/x) to describe the transfer, or frequency response function of y to x, where y and x are linearly related variables. Thus, if

$$y_t = \alpha x_t + \beta u_t + \dots \tag{9}$$

then (y/x) is the function $\alpha(z)$, in discrete time, or the function $\alpha(\omega)$, in continuous time.

10.2 Some Examples

The following is a typical regulation problem, fairly simple, although not the simplest. Let y_t denote the azimuthal setting of a radar aerial, as a function of time, and suppose that one wishes y_t to follow a stationary input signal u_t as closely as possible. If the aerial is driven by a motor exerting torque x_t, then we shall suppose that there are linear relations

$$\alpha y = x \tag{1}$$

$$x = \beta^{(1)} y + \beta^{(2)} u \tag{2}$$

Here α, $\beta^{(1)}$ and $\beta^{(2)}$ are operators, to be understood in the sense of section (1). The operator α is given, and represents the response of the aerial to the motor; one might expect something of the type

$$\alpha = MD^2 + KD \tag{3}$$

where M and K represent the angular inertia and the damping of the aerial. The operators $\beta^{(1)}$ and $\beta^{(2)}$, which represent the response of the motor to input and the actual aerial setting, are at least partially at our disposal. They are to be chosen so as to minimise the deviation $y_t - u_t$ in some sense.

Seeing that the variables y_t and x_t are rigidly related, it is plain that the controlling relation (2) is not uniquely determined, in the sense that there are several choices of $\beta^{(1)}$ and $\beta^{(2)}$ which give the same final relations between x, y and u. For example, if we eliminate y from relations (1) and (2), then we obtain formally

$$x = \left(\frac{\alpha\beta^{(2)}}{\alpha - \beta^{(1)}}\right)u = \beta u \tag{4}$$

say, and this could be regarded as a modification of the controlling relation (2), which is equivalent to it. In fact, it very often is not, for reasons which we shall discuss later in the section. Equation (4) describes the relation which *holds* between x and u, and is useful for this reason, but in general it will be something of type (2) rather than (4), which one uses to *determine* x at any instant.

If we chose the operators $\beta^{(1)}$ and $\beta^{(2)}$ so that β were equal to α, then we should get perfect following. However, this will rarely be practicable. The motor will have limited power, and a torque $x = \alpha u$ may not be realisable, since the variable αu may become excessively large; even infinite. In general, one would need a motor of infinite power and instantaneous response if the inertial system (1), (3) were to follow an arbitrary input perfectly, even though this input were known in advance.

If one is using mean-square criteria of performance, as we shall do, then, in view of what has been said, it seems reasonable to optimise the system (1), (2) by minimising $E(y_t - u_t)^2$ subject to a restriction on $E(x_t^2)$. That is, $\beta^{(1)}$ and $\beta^{(2)}$ are to be chosen so as to minimise

$$V = E[(y_t - u_t)^2 + \lambda x_t^2] \tag{5}$$

where λ can either be interpreted as a Lagrangian multiplier (to be chosen finally so that the rate of working, $E(x_t^2)$, shall have a definite value), or else simply as a coefficient which indicates the relative weight which one attaches to large values of $y - u$ or of x. Thus, the smaller λ, the less the relative penalty attached to a heavy rate of working of the motor, and so the more faithfully y will follow u in the optimised system.

Similarly, if there are practical restrictions on the rate of change of torque, due to factors such as inertia of the motor rotor, then the criterion function might be modified to

$$V = E[(y_t - u_t)^2 + \lambda_1 x_t^2 + \lambda_2 (Dx_t)^2] \tag{6}$$

If the whole course of u_t is known in advance, then the linear operator β, defined in (4), can be chosen without restriction. However, if at time t the value of u is known only up to time $t - h$, then β must be of the form

$$\beta u_t = \int_h^{\infty} \beta_s u_{t-s}\, ds \tag{7}$$

and the optimal β_s will be the solution of a Wiener–Hopf integral equation.

Let us now return to the distinction between the determination of the control variable x from (2) or from (4). If it is determined from (4), then the system is what a servo-engineer would term an *open-loop* system. If it is determined from (2), with $\beta^{(1)} \neq 0$, then there is *feedback*, and one has a *closed-loop* system.

Closed-loop operation is essential if there are disturbances; if, for example, relation (1) were modified to

$$\alpha y_t = x_t + \varepsilon_t \tag{8}$$

where $\{\varepsilon_t\}$ is a process of injected error (corresponding, perhaps, to deflection of the aerial by wind gusts). We shall see in section (4) that if there is such error then the optimised values of $\beta^{(1)}$ and $\beta^{(2)}$ immediately become determinate: the optimisation problem has a unique solution.

However, there is yet another reason for requiring feedback. If (1) and (4) were the two relations determining x and y, then y would be related to u by

$$y = \alpha^{-1}\beta u \tag{9}$$

Suppose, however, that α is singular (i.e. has a non-unique inverse) as is actually the case in (3). Then y as given by (9) is indeterminate. For example, if α is given by (3), then there is an arbitrary constant in y, which means that if y is out of correspondence with u initially, then no choice of β will bring it into correspondence.

On the other hand, if we solve (1), (2) for y we obtain

$$y = \frac{\beta^{(2)}}{\alpha - \beta^{(1)}} u \tag{10}$$

and if $\beta^{(1)}$ is such that $\alpha - \beta^{(1)}$ is non-singular, then this solution is determinate. In general, the control relations must be such that the square matrix of operators determining output and control variables (y, x) in terms of input variables (u) is non-singular.

As a second example, of much the same type, but more specific: suppose we wish a sequence $\{y_t\}$ to follow a given sequence $\{u_t\}$ closely, but without changing too rapidly from instant to instant. It then seems reasonable to choose y_t as the linear function of currently known values of u which minimises

$$V = E[(y_t - u_t)^2 + \lambda(y_t - y_{t-1})^2] \tag{11}$$

As before, λ is a measure of the seriousness of "discontinuity" relative to "deviation". Since y is to be determined directly in terms of u, there being no intermediate x variate, the question of operator inversion does not arise.

Our third example is a simplified version of a situation considered by Holt *et al.* (1960); it concerns the planning of factory production of a single type of article. The model is formulated in discrete time. For the tth time interval we define the variables

P_t = production
W_t = work-force
I_t = inventory (stock) at the end of period
S_t = orders during period

The *cost function* to be minimised is then defined as

$$V = E[A_1(W_t - W_{t-1})^2 + A_2(P_t - W_t)^2 + A_3(I_t - B)^2] \quad (12)$$

if P and W are expressed in appropriate units. The first term represents hiring and laying-off costs caused by variations in work force, the second overtime or idle-time costs incurred if the size of work force is not correct for the current rate of production, while the third represents storage or supply costs incurred if stock is respectively too large or too small.

It is assumed that all orders are satisfied, a negative inventory corresponding to supply of goods from some other source (at extra cost). There is thus a relation

$$P_t - S_t = I_t - I_{t-1} \quad (13)$$

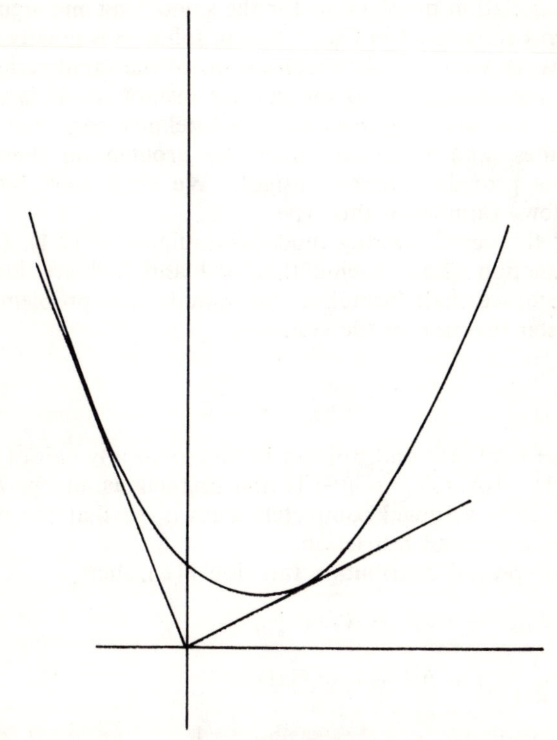

An illustration of how a quadratic cost function is used to approximate a piecewise linear cost function.

One could thus substitute for P in (12):

$$V = E[A_1(W_t - W_{t-1})^2 + A_2(S_t + I_t - I_{t-1} - W_t)^2 + A_3(I_t - B)^2] \quad (14)$$

and determine I and W (and hence also P) linearly in terms of the currently known ζ in such a way as to minimise expression (14).

This is a case for which it is quite evident that a quadratic criterion function is hardly realistic. For example, consider the last term in (12), relating to inventory. The cost of positive inventory, storage, is quite different in nature and magnitude from the cost of negative inventory, which is a penalty incurred by having to satisfy orders from outside sources. Both costs probably depend linearly rather than quadratically upon I, although with very different coefficients, as illustrated in the figure. The quadratic function $A_3(I - B)^2$ can only represent this broken cost-function approximately in the probable region of operation. The constant B gives the function some degree of asymmetry.

10.3 Regulation when the Input Series Have Known Future

For the regulation problem, as for the smoothing and signal extraction problems considered in Chs. 5 and 6, solution is greatly simplified if one can work with complete realisations of the input series, instead of just the realisations up to the current instant, t. It is seldom in practice that one will have this complete foreknowledge, but occasionally one does, and the solution of the problem in these circumstances does provide valuable insight. We shall start by working through a few examples of this type.

Consider the aerial steering model formulated in (2.1), (2.2), with criterion function (2.5). Seeing that $\beta^{(1)}$ and $\beta^{(2)}$ are individually indeterminate, we shall formulate the optimisation problem in terms of the transfer function of the system:

$$\theta = \left(\frac{y}{u}\right) = \frac{\beta^{(2)}}{\alpha - \beta^{(1)}} \quad (1)$$

Once this is found, $\beta^{(1)}$ and $\beta^{(2)}$ can be chosen as any pair of functions satisfying (1), provided $\alpha - \beta^{(1)}$ is non-singular as an operator. The input series u_t is assumed completely known, so that the operator θ can be chosen without restriction.

If $\{u_t\}$ has spectral distribution function $F(\omega)$, then

$$\begin{aligned} V &= E[((\theta - 1)u)^2 + \lambda(\alpha\theta u)^2] \\ &= \frac{1}{2\pi}\int [|1 - \theta|^2 + \lambda|\alpha\theta|^2]dF(\omega) \\ &= \frac{1}{2\pi}\int \left[(1 + \lambda|\alpha|^2)\left|\theta - \frac{1}{1 + \lambda|\alpha|^2}\right|^2 + \frac{\lambda|\alpha|^2}{1 + \lambda|\alpha|^2}\right]dF(\omega) \quad (2) \end{aligned}$$

It is thus evident that for optimum following

$$\theta = \frac{1}{1 + \lambda|\alpha|^2} \tag{3}$$

$$V = \frac{\lambda}{2\pi}\int \frac{|\alpha|^2\, dF(\omega)}{1 + \lambda|\alpha|^2} \tag{4}$$

Ex. 1. Show that if $\alpha = MD^2$, i.e. $\alpha(\omega) = -M\omega^2$, then under optimal following

$$y_t = \frac{1}{2\lambda}\int_0^\infty e^{-fs/\sqrt{2}} \sin\left(\frac{fs}{\sqrt{2}} + \frac{\pi}{4}\right) [u_{t+s} + u_{t-s}]\, ds$$

and that if

$$F'(\omega) = \frac{\sigma^2}{\omega^2 + \mu^2}$$

then

$$V = \frac{\sigma^2}{2\sqrt{2}f[f^2 + \mu^2 + \sqrt{2}f\mu]}$$

where

$$f = (\lambda M^2)^{-1/4}$$

Ex. 2. Carry through the same calculation in the case $\alpha = MD^2 + KD$.

Although we have spoken of aerial steering, for the sake of concreteness, the example is obviously a fairly general one.

The filter with frequency response (3) is not physically realisable, since it is not the transform of a one-sided weight-function. In fact, $\theta(\omega)$ is symmetric in ω, so that y is always symmetric in its dependence upon the future and past of u.

Of especial interest are the cases of small and large λ, for which $\beta = (x/u)$ takes the respective approximate forms,

$$\beta(\omega) \sim \alpha(\omega) \tag{5}$$

$$\beta(\omega) \sim (\lambda\alpha(-\omega))^{-1} \tag{6}$$

for fixed ω. These correspond respectively to the case of large and small motor power. It is not too fanciful to term these extreme cases the "brute-force" and "jiu-jitsu" methods of steering. For, recall that $R(\omega) = \alpha(\omega)^{-1}$ is the frequency response function of y to x; typically this will tend to zero as ω becomes large, corresponding to the fact that a heavy aerial will have more difficulty in following the higher frequency components of input. Now, the filter (5) has the transfer function inverse to R, and corresponds to an attempt to completely compensate for aerial response. If $|R(\omega)|$ is small then $|\beta(\omega)|$ is large: the pre-filter attempts to compensate for the inertia of the aerial by exaggerating just those frequencies in the input which the aerial has difficulty in following —these will usually be the higher ones.

For case (6), however, we see that $|\beta(\omega)|$ is proportional to $|R(\omega)|$; the frequencies which are relatively exaggerated now are those which the aerial can follow easily; the difficult high-frequency motions are not even attempted.

We shall not consider in this section the more general case formulated in (2.2), (2.8), since it would be unrealistic to assume that the error process $\{\varepsilon_t\}$ had a perfectly predictable future.

The expression for θ in (3) will hold also for the case of discrete time, if the conventions of section (1) concerning notation are observed.

Ex. 3. Consider the second example of the last section, and show that, if y may depend upon the whole realisation of u, then expression (11) is minimised when

$$y_t = \frac{1}{1 + \lambda(1 - U)(1 - U^{-1})} u_t = \left(\frac{1 - \xi}{1 + \xi}\right) \sum_{-\infty}^{\infty} \xi^{|j|} u_{t-j}$$

where

$$\xi = \frac{(1 + 2\lambda) - \sqrt{1 + 4\lambda}}{2\lambda}$$

Why must the operator acting on u_t be expanded in just the manner indicated?

Finally, consider the production example, with cost-function (2.12). A point of interest of this example is that there are two transfer functions to be determined, namely, those giving the dependence of W and of I on S. If these are denoted by β and γ respectively, so that, for example,

$$W_t = \beta(U)S_t = \sum_j \beta_j S_{t-j} \tag{7}$$

then

$$V = \mathscr{A}\{[A_1|1 - z|^2|\beta|^2 + A_2|1 + (1 - z)\gamma - \beta|^2 + A_3|\gamma|^2]|\psi_S|^2\} \tag{8}$$

where $|\psi_S(z)|^2$ is the spectral density function of the S_t sequence, in its canonically factorised form.

Now, either by completion of the square, as in equation (2), or by equation of differentials to zero, we find that the β and γ which unconditionally minimise expression (8) obey the equation system

$$\begin{bmatrix} A_1|1 - z|^2 + A_2 & -A_2(1 - z) \\ -A_2(1 - z^{-1}) & A_2|1 - z|^2 + A_3 \end{bmatrix} \begin{bmatrix} \beta \\ \gamma \end{bmatrix} = A_2 \begin{bmatrix} 1 \\ -1 + z^{-1} \end{bmatrix} \tag{9}$$

We thus find the optimum transfer functions to be

$$\left(\frac{W}{S}\right) = \beta = \left[1 + \frac{A_1}{A_2}|1 - z|^2 + \frac{A_1}{A_3}|1 - z|^4\right]^{-1} = (1 + Hf^2 + Gf^4)^{-1} \tag{10}$$

$$\left(\frac{I}{S}\right) = \gamma = -\frac{A_1}{A_3}(1 - z^{-1})^2(1 - z)\beta = \frac{iGf^3}{z^{\frac{1}{2}}}\beta \tag{11}$$

where, following Holt and his co-authors, we have introduced the convenient notations

$$H = A_1/A_2, \quad G = A_1/A_3 \tag{12}$$

$$f = \frac{z^{-\frac{1}{2}} - z^{\frac{1}{2}}}{i} = 2 \sin\left(\frac{\omega}{2}\right) \tag{13}$$

(as before, $z = e^{-i\omega}$).

From (2.13) and (11) we find

$$\left(\frac{P}{S}\right) = 1 + (1 - z)\gamma = (1 + Hf^2)\beta \tag{14}$$

As the frequency ω increases from 0 to π, f increases also, from 0 to 2, so that equations (10), (11), (14) give one some idea of the sensitivity of the W, I and P sequences to different frequency components in the order sequence, S. Relation (10) shows that response in work-force will fall off rather quickly with increasing frequency. The additional f^3 factor in (11) shows that inventory is relatively sensitive to higher frequencies, i.e. to sudden changes, in orders. Production itself is intermediate in response.

Of direct importance are the actual components in the cost function. These components have transfer functions whose moduli are

$$\left.\begin{aligned} \left|\frac{I}{S}\right| &= G|f|^3\beta \\ \left|\frac{W_t - W_{t-1}}{S}\right| &= |f|\beta \\ \left|\frac{P - W}{S}\right| &= Hf^2\beta \end{aligned}\right\} \tag{15}$$

Inspecting the various powers of f which occur, we see that it is inventory costs which are sensitive to rapid variations in orders, costs in changes of work-force which are sensitive to slow variations, while costs due to incorrect size of work-force are intermediate in response.

Ex. 4. Show that in formula (7)

$$\beta_j = \beta_{-j} = \frac{(1-\xi)^2(1-\eta)^2}{(1-\xi\eta)(\xi-\eta)}\left[\frac{\xi^{j+1}}{1-\xi^2} - \frac{\eta^{j+1}}{1-\eta^2}\right] \quad (j \geqslant 0)$$

where ξ, η are the two roots of

$$1 + H(1-z)(1-z^{-1}) + G(1-z)^2(1-z^{-1})^2 = 0$$

which lie inside the unit circle. Show that ξ, η are real if $H^2 \geqslant 4G$, but that otherwise they form a conjugate pair, $e^{-\lambda \pm i\mu}$, say, for which

$$\cosh(\lambda)\cos(\mu) = 1 + \frac{H}{4G}$$

10.4 Statistical Averages, Time Averages and Non-stationary Inputs

The criterion function (2.5) can be otherwise written as

$$V = E[(y_t - u_t)^2 + \lambda(\alpha y_t)^2] \tag{1}$$

and the transfer function (3.1) minimising this is given by (3.3), which does not depend upon the statistical properties of the input process $\{u_t\}$ at all. Thus, the relation (3.3) is the optimum one for any stationary input: one wonders whether it is also optimum for other inputs.

If u_t is a known function, and there are no random disturbances in the system, then it might seem more natural to attempt to minimise a time-average

$$V'_{mn} = \frac{1}{n-m+1}\sum_m^n [(y_t - u_t)^2 + \lambda(\alpha y_t)^2] \tag{2}$$

(the sum being replaced by an integral, in continuous time) rather than the stochastic average (1). For one thing, one can then consider the optimisation for a definite function u_t, rather than for a class of functions and, for another, the criterion (2) does not presume stationarity of $\{u_t\}$ as (1) does.

In fact, the two approaches give equivalent results in a large class of cases. Suppose we consider discrete time, and suppose that α is a one-sided operator:

$$\alpha y_t = \sum_0^\infty \alpha_j y_{t-j} \tag{3}$$

Then, differentiating (2) with respect to y_t, we find

$$y_t - u_t + \lambda \sum_{s=m}^{n} \alpha_{s-t}\Big(\sum_{j=0}^{\infty} \alpha_j y_{s-j}\Big) = 0 \qquad (t = m, m+1, \ldots n) \tag{4}$$

or, recalling the notation (1.4),

$$(1 + \lambda|\alpha|^2)y_t = u_t \quad (t = m, m+1, \ldots n) \tag{5}$$

provided that

$$\sum_{s=n+1}^{\infty} \alpha_{s-t}\Big(\sum_{j=0}^{\infty} \alpha_j y_{s-j}\Big) = 0 \qquad (t = m, m+1, \ldots n) \tag{6}$$

Relation (5) can be regarded as a difference equation, to be solved subject to the boundary conditions (6) and the prescription of y_t, $(t < m)$. Relations (3.1) and (3.3) can be regarded as the *same* difference equation, now holding for all t and with solution subject to regularity conditions which ensure that $\{y\}$ is stationary if $\{u\}$ is. The question is, whether the solution of the former system tends to that of the latter as $m \longrightarrow -\infty$, $n \longrightarrow \infty$. The answer will depend upon the rate of growth of u_t as $|t| \longrightarrow \infty$, and upon the location of the singularities of $\theta(z)$. We shall not consider the question further, except to state that the two systems will become equivalent, with probability one, if $\alpha(z)$ is a polynomial and u_t a realisation of a stationary process.

Ex. 1. Consider the case $\alpha = 1 - U$, written out in (2.11), and treated on statistical lines in Ex. (3.3). Show that the time average (2) is minimised in this case by

$$y_t = \sum_{s=m}^{n} b_{ts}u_s + \lambda b_{tm}y_{m-1}$$

where

$$b_{ts} = \frac{1-\xi}{1+\xi}\left[\xi^{|t-s|} + \frac{\xi^{m+n-t-s-1} - \xi^{n-m-t+s+1} - \xi^{-m-n+t+s} - \xi^{-m+n+t-s+1}}{\xi^{n-m+1} + \xi^{m-n-2}}\right]$$

Compare with the solution of Ex. (3.3).

Even with stationary inputs, there are sometimes good reasons for working with the time average criterion (2) rather than the statistical average criterion (1). For example, the process may have a history of non-optimal running, so that the first attempts at optimisation begin at $t = m$. A certain time will elapse before the process reaches a stationary state under the new rules, and one may wish to optimise running during the transition period. As another example, one may not contemplate indefinite running of the process, but only intend to work to a "time-horizon" at $t = n$. It is interesting that if one works with a receding time-horizon, $n = t + \nu$, the regulation rules calculated on the basis of relatively small ν often do not differ greatly from those calculated on the basis of indefinite operation ($\nu = \infty$).

If there is any uncertainty in the future inputs, that is, either in u or in any injected disturbances, then one must obviously work to some extent with statistical averages. Nevertheless, we shall see in section (8) that there are "certainty equivalence" principles which allow one to still think in terms of minimising a known quadratic form such as (2), even though some of the variables in the quadratic may be unknown random variables.

There may be non-stationary inputs u, for which αy and $y - u$ are themselves stationary, so that evaluation of performance by the statistical average (1) is meaningful. Let us suppose that u is an accumulated process of the type considered in section (8.5), so that u is derived from a stationary process $\{v_t\}$ by a relation of the type

$$(1 - \rho U)^p u_t = v_t \tag{7}$$

(discrete time) or

$$\left[\frac{d}{dt} + (1-\rho)\right]^p u_t = v_t \tag{8}$$

(continuous time) where we consider the limit $\rho \uparrow 1$. In either case, for $0 < \rho < 1$ the process u_t will be stationary, with a proper s.d.f. $f_{uu}(\omega)$, but, as $\rho \uparrow 1$, $f_{uu}(\omega)$ will develop a singularity of the type ω^{-2p}. Now, we know from equation (3.3) that

$$\left(\frac{y-u}{u}\right) = \frac{-\lambda|\alpha|^2}{1+\lambda|\alpha|^2} \tag{9}$$

$$\left(\frac{x}{u}\right) = \left(\frac{\alpha y}{u}\right) = \frac{\alpha}{1+\lambda|\alpha|^2} \tag{10}$$

and if both $y - u$ and αy are to be stationary it is necessary that these transfer functions should be $O(\omega^p)$ for small ω. This will be true if and only if $\alpha(\omega)$ (or $\alpha(z)$) is $O(\omega^p)$ at the origin. That is, the optimum filter will handle a pth order accumulated process (optimally) if, and only if, the given operator α behaves as a differential (or difference) of at least order p. So, with the operator (2.3), the system could certainly handle a simple accumulated process, and could handle a doubly accumulated process if $K = 0$ (zero damping).

If $\alpha(\omega)$ is $O(\omega^p)$ at the origin, then we see from (3.3) that

$$\left(\frac{y}{u}\right) = 1 + O(\omega^{2p}) \tag{11}$$

That is (cf. Theorem 8.5.2) the following system is exact for inputs $u_t = t^j$ ($j = 0, 1, 2, \ldots 2p - 1$) and so the first $2p$ dynamic lags are zero (see James, Nichols and Phillips (1947) for definition).

10.5 Optimal Regulation for Systems with Inputs of Uncertain Future

The possible variety of systems is infinite: this is plain when one considers that there may be several inputs, predictable or not, that error may be injected into the system at many points, that there may be several outputs, in the sense that the m.s. criterion is a function of variables at several points of the system, and then there may be built-in elements, such as lags, which are intrinsic and cannot be changed. For this reason, we shall not attempt a very general treatment, but rather consider a succession of representative cases, each of fair generality in itself.

The methods of attack should in this way be made plain enough that the reader can apply them to other problems. In section (7) we shall work through a number of specific examples.

The principal tools are just those used for treatment of the prediction problem; although some preliminary transformations are often needed before those tools can be employed. For simplicity and uniformity we shall consider problems in discrete time in this section: solutions to the continuous time equivalents follow immediately by analogy.

(i) *Undisturbed following*

Let us reconsider the following problem solved in equations (3.1)–(3.4), but let us now suppose that at time t the only values of u_s known are those for $s \leqslant t - h$ (where h may be positive or negative), so that the transfer function (3.1) is now restricted to being of the form

$$\theta(z) = \sum_{h}^{\infty} \theta_j z^j \tag{1}$$

We shall suppose for the moment that the $\{u_t\}$ process is purely non-deterministic with s.d.f. having the canonical factorisation

$$f_{uu}(\omega) = g_{uu}(z) = \psi_u(z)\psi_u(z^{-1}) = |\psi_u|^2 \tag{2}$$

Then we can write the criterion function (2.5) in the form

$$V = \mathscr{A}[|1 - \theta|^2|\psi_u|^2 + \lambda|\alpha\theta\psi_u|^2] \tag{3}$$

This must now be minimised with respect to the coefficients θ_h, $\theta_{h+1}, \ldots$ Differentiating with respect to θ_j we obtain

$$\mathscr{A}[z^{-j}H(z) + z^j H(z^{-1})] = 0 \qquad (j \leqslant h) \tag{4}$$

where

$$H(z) = [(1 + \lambda|\alpha|^2)\theta - 1]|\psi_u|^2 \tag{5}$$

Equation (4) implies that the coefficient of z^j in the Laurent expansion of $H(z)$ on $|z| = 1$ is zero ($j \geqslant h$), so we must have

$$H(z) = \sum_{-\infty}^{h-1} \tag{6}$$

where by the sum we indicate an unknown function whose Laurent expansion on $|z| = 1$ includes only powers z^j in the range indicated. From (5) and (6) it readily follows that

$$[(1 + \lambda|\alpha|^2)\theta - 1]\psi_u = \sum_{-\infty}^{h-1} \tag{7}$$

and thence, by the Wiener–Hopf argument described in Ch. 6, that

$$\theta = \frac{1}{P\psi_u}\left[\frac{\psi_u}{\bar{P}}\right]_h \tag{8}$$

where $P(z)$ is determined by the canonical factorisation

$$1 + \lambda|\alpha|^2 = |P|^2 \tag{9}$$

As h tends to $-\infty$, solution (8) tends to the unrestricted solution (3.3). Note that θ now depends on the s.d.f. of the input process.

We shall consider the case where the input is partly deterministic in the next section.

Ex. 1. Show that if the criterion function can be written

$$V = E[[(A\theta - B)u]^2 + (Cu)^2] \tag{10}$$

where A, B and C are operators, and A is canonical (in the sense that $A(z)$ is analytic and zero-free in $|z| \leqslant 1$), then the optimal transfer function is

$$\theta(z) = \frac{1}{A(z)\psi_u(z)}[B(z)\psi_u(z)]_h \tag{11}$$

Ex. 2. Write down the equivalents of (8) and (11) in continuous time.

(ii) *Disturbed following*

Consider now the disturbed system

$$\alpha y_t = x_t + \varepsilon_t \tag{12}$$

$$x_t = \beta^{(1)}y_t + \beta^{(2)}u_t \tag{13}$$

with, again, the criterion function (2.5). Here x_t is, as before, the signal or control variable, whose effect is now perverted by the error ε_t. We shall see that the presence of an error will make the optimisation determinate, so that $\beta^{(1)}$ and $\beta^{(2)}$ will be determined individually.

Suppose we restrict these two operators to being of the forms

$$\beta^{(1)}(z) = \sum_k^\infty \beta_j^{(1)} z^j \tag{14}$$

$$\beta^{(2)}(z) = \sum_h^\infty \beta_j^{(2)} z^j \tag{15}$$

on the assumption that there is a lag h in information on u, and a lag k in observation and reporting of the value of the controlled variable, y.

We shall assume that the stationary processes $\{u\}$ and $\{\varepsilon\}$ are purely non-deterministic and mutually uncorrelated, with s.d.f.'s $|\psi_u|^2$ and $|\psi_\varepsilon|^2$.

Solving the system (12), (13), we have

$$y = \frac{\beta^{(2)}u + \varepsilon}{\alpha - \beta^{(1)}} \tag{16}$$

$$x = \frac{\alpha\beta^{(2)}u + \beta^{(1)}\varepsilon}{\alpha - \beta^{(1)}} \tag{17}$$

Instead of using $\beta^{(1)}$ and $\beta^{(2)}$, it is convenient to work in terms of the transfer functions

$$\theta = \left(\frac{y}{u}\right) = \frac{\beta^{(2)}}{\alpha - \beta^{(1)}} \tag{18}$$

$$\phi = \left(\frac{y}{\varepsilon}\right) = \frac{1}{\alpha - \beta^{(1)}} \tag{19}$$

in terms of which

$$\begin{aligned} V &= E\{[\theta - 1)u + \phi\varepsilon]^2 + \lambda[\alpha\theta u + (\alpha\phi - 1)\varepsilon]^2\} \\ &= \mathscr{A}\{[|1 - \theta|^2 + \lambda|\alpha\theta|^2]|\psi_u|^2 + [|\phi|^2 + \lambda|1 - \alpha\phi|^2]|\psi_\varepsilon|^2\} \end{aligned} \tag{20}$$

The functions θ and ϕ will have expansions on the unit circle

$$\theta(z) = \sum_h^\infty \theta_j z^j \tag{21}$$

$$\phi(z) = \sum_0^\infty \phi_j z^j \tag{22}$$

All the coefficients θ_j are at our disposal, and we can optimise freely with respect to them. Of the ϕ_j coefficients only $\phi_k, \phi_{k+1}, \ldots$ are disposable, however: the coefficients $\phi_0, \phi_1, \ldots \phi_{k-1}$ are equal to the corresponding coefficients in a formal Taylor expansion of α^{-1}, as is clear from (19) and (14).

Upon a comparison of (20) and (21) with (1) and (3) it is apparent that the optimum θ will be just that given before, in equation (8). Differentiating (20) with respect to the disposable ϕ_j we find, as in (7), that

$$[(1 + \lambda|\alpha|^2)\phi - \lambda\bar{\alpha}]\psi_\varepsilon = \sum_{-\infty}^{k-1} \tag{23}$$

so that

$$P\psi_\varepsilon\phi = \lambda\left[\frac{\bar{\alpha}\psi_\varepsilon}{\bar{P}}\right]_k + \sum_{0}^{k-1} \tag{24}$$

where the sum represents a polynomial, as yet undetermined. However, the polynomial must be equal to

$$\pi(z) = [P\psi_\varepsilon\phi]^{(k-1)} \tag{25}$$

which is known, because $\phi_0, \phi_1, \ldots \phi_{k-1}$ are known. In fact, from what was said concerning those coefficients, we know that $\pi(z)$ is equal to the sum of the first k terms in a formal Taylor expansion of $P\psi_\varepsilon/\alpha$, something which we shall indicate by

$$\pi(z) = \left[\frac{P\psi_\varepsilon}{\alpha}\right]^{(k-1)*} \tag{26}$$

the asterisk emphasising that the expansion is not to be taken on $|z| = 1$, but around $z = 0$.

We thus derive a final solution for ϕ:

$$\phi = \frac{1}{P\psi_\varepsilon}\left[\frac{P\psi_\varepsilon}{\alpha}\right]^{(k-1)*} + \frac{\lambda}{P\psi_\varepsilon}\left[\frac{\bar{\alpha}\psi_\varepsilon}{\bar{P}}\right]_k \tag{27}$$

If $\alpha(z)$ has no singularities in $|z| \leqslant 1$, then we can drop the asterisk in (27), and substitute for $\lambda\bar{\alpha}$ from (9); solution (27) then reduces to

$$\phi = \frac{1}{\alpha} - \frac{1}{P\psi_\varepsilon}\left[\frac{\psi_\varepsilon}{\bar{P}\alpha}\right]_k \tag{28}$$

However, in general this is not permissible, and the solution must be taken from (27).

Ex. 3. Show that if $\{\varepsilon_t\}$ is generated by a moving average scheme of order less than k, then the second term in the right-hand member of (27) is zero, and

$$\beta^{(1)} = \alpha - P\psi_\varepsilon/[P\psi_\varepsilon/\alpha]^{(k-1)*} \tag{29}$$

Ex. 4. Consider a *pure stabilisation process*, in which one merely wishes to hold y_t on a constant value (zero, say), so that $u_t \equiv 0$ in the scheme (12), (13), (2.5). Show that for optimum stabilisation $\beta^{(1)}$ is still determined by equation (27), and $\beta^{(2)}$ may be taken equal to zero.

Ex. 5. Modify the above treatment for the case where there are errors in observation of y_t and u_t, so that while relations (12) and (2.5) remain unchanged, in (13) we should have to substitute $y_t + \eta_t^{(1)}$ and $u_t + \eta_t^{(2)}$ for y_t and u_t, the η's being stationary error processes.

Ex. 6. Consider the case in which there is an error in the formation of the signal sequence, so that relation (13) must be modified to

$$x_t = \beta^{(0)}x_t + \beta^{(1)}y_t + \beta^{(2)}u_t + \varepsilon'_t$$

where

$$\beta^{(0)}(z) = \sum_{1}^{\infty} \beta_j^{(0)} z^j$$

is to be determined, and ε'_t is a stationary error process with known s.d.f. Determine the optimum set of operators β. (If $\varepsilon'_t \equiv 0$, then the problem would be indeterminate, but otherwise there will be a method of forming the control sequence which has greatest "noise-immunity".)

10.6 Inputs with Deterministic and Non-stationary Components

We have been assuming that all series have zero mean. This will scarcely ever be true, and, indeed, the criterion function corresponding to (2.5) will in general be as much dependent on the means of the variables as on their fluctuations. However, the restriction is easily remedied.

Let us suppose that the input series u_t is now replaced by $\bar{u}_t + u_t$, where $\bar{u}_t$ is a sequence whose whole realisation is known in advance, while u_t is a purely non-deterministic stationary process of the type already considered, whose mean (zero) and covariances are functionally independent of $\bar{u}_t$. The case $\bar{u}_t =$ const. corresponds to the presence of a non-zero constant mean.

Suppose that the control and output variables are similarly decomposed into a deterministic mean, and a purely non-deterministic fluctuation: $\bar{x}_t + x_t$ and $\bar{y}_t + y_t$. Suppose that the criterion function is

$$V = E[Q(\bar{y}_t + y_t, \bar{x}_t + x_t, \bar{u}_t + u_t)] \quad (1)$$

where

$$Q(\xi, \eta, \zeta) = Q_0 + Q_1(\xi, \eta, \zeta) + Q_2(\xi, \eta, \zeta) \quad (2)$$

is a second-degree function of its arguments; Q_0, Q_1 and Q_2 constituting respectively the constant, linear and quadratic components. If $\bar{u}_t$ is a realisation of a stationary process, then the expectation in (1) is over realisations of $\bar{u}_t$ as well as over values of the fluctuations u_t (and ε_t): this, for simplicity, is the viewpoint we shall take.

One can also adopt the viewpoint that $\bar{u}_t$ is a definite function of time, and include a time-averaging as well as a statistical averaging in the criterion function. If $\bar{u}_t$ is not too "non-stationary" in form this approach will give equivalent results—see section (4).

Now, by expanding Q in (1), and remembering that the fluctuations have zero expectation, we find that

$$V = E[Q(\bar{y}_t, \bar{x}_t, \bar{u}_t)] + E[Q_2(y_t, x_t, u_t)] \quad (3)$$

That is, if $\bar{u}_t$ and u_t can be separated in the input, then the following of the two components can be considered quite separately. The regulation of fluctuations is achieved by the methods of section (5), and based on the criterion function $E[Q_2(y, x, u)]$. The optimal following of the deterministic component of input is achieved by the methods of section (3), based on the criterion function $E[Q(\bar{y}, \bar{x}, \bar{u})]$.

This situation is particularly simple if $\bar{u}_t$ is a constant, when one simply chooses constants $\bar{x}$ and $\bar{y}$, subject to relations between the two sequences, which minimise $Q(\bar{y}, \bar{x}, \bar{u})$.

As an example, let us consider the disturbed following problem again. If we use U_t to denote total input

$$U_t = \bar{u}_t + u_t \tag{4}$$

etc. then we assume

$$V = E[(Y_t - U_t)^2 + \lambda X_t^2] \tag{5}$$

If we assume that the error sequence ε_t is still purely non-deterministic (although we could well cope with the more general case), then $\bar{y}_t$ will be related to $\bar{u}_t$ by the operator (3.3), while the relation of y_t to its own past and to u_t will be given by the treatment of section (5). More explicitly, corresponding to (5.12) and (5.13) we shall have relations

$$\alpha Y = X + \varepsilon \tag{6}$$

$$X = \beta^{(1)} Y + \beta^{(2)} u + \beta^{(3)} \bar{u} \tag{7}$$

which imply

$$\left.\begin{aligned} \alpha y &= x + \varepsilon \\ x &= \beta^{(1)} y + \beta^{(2)} u \end{aligned}\right] \tag{8}$$

$$\left.\begin{aligned} \alpha \bar{y} &= \bar{x} \\ \bar{x} &= \beta^{(1)} \bar{y} + \beta^{(3)} \bar{u} \end{aligned}\right] \tag{9}$$

The operators $\beta^{(1)}$ and $\beta^{(2)}$ in system (8) will again be determined from (5.8) and (5.27). The operator $\beta^{(3)}$ in (9) must then be chosen so that

$$\bar{y} = \frac{1}{1 + \lambda|\alpha|^2} \bar{u} \tag{10}$$

i.e.

$$\beta^{(3)} = \frac{\alpha - \beta^{(1)}}{1 + \lambda|\alpha|^2} = \frac{1}{\phi|P|^2} \tag{11}$$

Ex. 1. Suppose one is working in continuous time, and one wishes to stabilise y on to a constant value μ, so that

$$V = E[(y - \mu)^2 + \lambda x^2]$$

Show that the optimum control signal is

$$\begin{aligned} x_t &= \frac{\alpha(0) - \beta^{(1)}(0)}{1 + \lambda[\alpha(0)]^2} \mu + \beta^{(1)} y_t \\ &= \frac{\alpha(0) + \lambda\alpha(0)^2 \beta^{(1)}(0)}{1 + \lambda\alpha(0)^2} \mu + \beta^{(1)}(y_t - \mu) \end{aligned}$$

where $\beta^{(1)}$ is determined by the continuous time equivalent of (5.28).

Of course, just as much as in prediction, a deterministic component may be known only to the extent that it can be represented in the form

$$\bar{u}_t = \sum b_j g_j(t) \tag{12}$$

where the $g_j(t)$ are known functions, and the b_j unknown coefficients. Just as in the case of prediction, the b_j could be estimated from the sample (see sections (4.3), (8.3)), although the calculation would not be one that could be mechanised naturally.

A more natural approach is to again consider the possibility that the input u_t may be an accumulated process, as defined in section (8.5). Just as in section (4), for satisfactory operation under these conditions one would demand that the two variates of the criterion function, $y_t - u_t$ and x_t, constituted stationary processes. We see from (5.16)–(5.19) that

$$\left(\frac{y-u}{u}\right) = \theta - 1 \tag{13}$$

$$\left(\frac{x}{u}\right) = \alpha\theta \tag{14}$$

Now, if $\psi_u(z)$ has a factor $(1 - \rho z)^{-p}$ and $\rho \uparrow 1$, then we see from (5.8) and Theorem (8.5.1) that

$$\begin{aligned}\theta(z) &= \frac{1}{1 + \lambda|\alpha|^2} + O[(1 - z)^p] \\ &= \frac{1}{1 + \lambda|\alpha|^2} + O(\omega^p)\end{aligned} \tag{15}$$

and we see from (13)–(15) that equations (4.9) and (4.10) are still true to within a term of order ω^p. We thus come to the same conclusion as in section (4); the system will operate properly with a p-fold accumulated input if and only if the response function α is $O(\omega^p)$ at the origin, i.e. if the operator α behaves as a difference or differential (in discrete or continuous time respectively) of order p at least. With this condition fulfilled, $\theta = 1 + O(\omega^p)$, so that at least the first p dynamic lags of the system are zero.

Ex. 2. Show that if $\alpha(z)$ has a factor $(z - z_0)^p$, where p is an integer, then

$$\theta(z) = 1 + O[(z - z_0)^p]$$

Show that this implies that the inputs $u_t = t^j z_0{}^{-t}$ $(j = 0, 1, \ldots p - 1)$ would ultimately be followed without systematic error.

Ex. 3. Derive the corresponding results for the case of continuous time.

10.7 Specific examples

(i) *A simple example of temperature stabilisation*

Let Y_t denote the temperature of an oven at time t, measured from room temperature, and let X_t denote the rate of fuel supply to the oven in the interval $(t - 1, t)$ measured in such units that we can write

$$Y_t = aY_{t-1} + X_t + \varepsilon_t \tag{1}$$

where ε_t is a random deviation. We are thus assuming a very simple situation, in which we can neglect heat-flow between different parts of the oven, and merely assume that temperature depends on past temperatures and heat supplies in the manner indicated by (1), a being a heat retention factor lying between 0 and 1.

Suppose we let X depend linearly upon past values of Y:

$$X_t = c + \beta(U)Y_t = c + \sum_1^\infty \beta_j Y_{t-j} \tag{2}$$

in such a way as to minimise

$$V = E[(Y_t - T)^2 + \lambda(X_t - \bar{x})^2 + 2\mu\bar{x}] \tag{3}$$

where $\bar{x}$ is the mean value of X and T the desired oven temperature. The terms in the cost function (3) are thus attributable respectively to temperature deviation, fuel supply variation, and average fuel rate.

This is then just a following (or stabilisation) problem of the type treated in sections (5) and (6), with the specialisations $\alpha(z) = 1 - az$, $k = 1$, and $U_t \equiv T$.

Now, in the notation (4) of section (6)

$$V = E(y^2 + \lambda x^2) + (\bar{y} - T)^2 + 2\mu\bar{x} \tag{4}$$

so that β is to be chosen to minimise $E(y^2 + \lambda x^2)$—just the problem considered in section 5(ii)—while c is to be chosen to minimise $(\bar{y} - T)^2 + 2\mu\bar{x}$. Now, from equations (1) and (2) we find that if $E(\varepsilon) = 0$, then

$$\left.\begin{aligned}(1 - a)\bar{y} &= \bar{x}\\ \bar{x} &= c + \beta(1)\bar{y}\end{aligned}\right] \tag{5}$$

Substituting for $\bar{x}$ from (5) into $(\bar{y} - T)^2 + 2\mu\bar{x}$, and minimising this expression with respect to $\bar{y}$, we find that under optimum running

$$\bar{y} = T - (1 - a)\mu \tag{6}$$

The corresponding values of $\bar{x}$ and c follow from equations (5): note that it is really c which is chosen to optimise the design, the values of $\bar{x}$ and $\bar{y}$ following automatically. It is interesting that $\bar{x}$ and $\bar{y}$ turn out to be independent of the operator β.

The optimum β will be identical with the $\beta^{(1)}$ determined by equations (5.19), (5.27). In the present case we have

$$\begin{aligned}|P|^2 = 1 + \lambda|\alpha|^2 &= 1 + \lambda(1 - az)(1 - az^{-1})\\ &= K^2(1 - \xi z)(1 - \xi z^{-1})\end{aligned} \tag{7}$$

say, where ξ is the zero of $1 + \lambda|\alpha|^2$ which lies inside the unit circle. Thus $P(z) = K(1 - \xi z)$.

Suppose that the process $\{\varepsilon_t\}$ is an uncorrelated one, so that $\psi_\varepsilon(z) = \text{const.} = \sigma$, say. Then, by (5.27),

$$\phi = \frac{1}{\alpha - \beta} = \frac{1}{\sigma P}\left[\frac{P\sigma}{\alpha}\right]^{(0)*} = \frac{1}{1 - \xi z} \tag{8}$$

so that

$$\alpha(z) - \beta(z) = 1 - \xi z \tag{9}$$

$$\beta(z) = (\xi - a)z \tag{10}$$

That is, as far as fluctuations from mean values are concerned, the regulation system is

$$x_t = (\xi - a)y_{t-1} \tag{11}$$

so that the relation

$$y_t - ay_{t-1} = x_t + \varepsilon_t \tag{12}$$

becomes

$$y_t - \xi y_{t-1} = \varepsilon_t \tag{13}$$

One finds readily that

$$\text{var } y = \frac{\sigma^2}{1 - \xi^2} \tag{14}$$

$$\text{var } x = \frac{\sigma^2(\xi - a)^2}{1 - \xi^2} \tag{15}$$

For a system which, when unregulated, is only just stable, quite a modest amount of regulation can bring about immense improvement. For example, consider the above case with $a = 0{\cdot}98$, $\sigma = 1$. For various values of $-\beta_1 = a - \xi$ (corresponding to the optimum solution for varying λ) we obtain the following results:

$-\beta_1$	0·00	0·08	0·18
ξ	0·98	0·90	0·80
var (y)	25·25	5·26	2·78
var (x)	0·000	0·034	0·090

The first case is the unregulated one, the second two correspond to increasing degrees of regulation. It will be seen that values of β_1 for which the value of var (x) is quite modest bring about a spectacular reduction in var (y).

For the case of temperature stabilisation one naturally has $0 < a < 1$, so the system is intrinsically stable. However, it is interesting to examine the general case, where a may conceivably lie outside the range $(-1, 1)$ and the unregulated system be unstable. We have

$$\xi = \frac{1 + \lambda(1 + a^2) - \sqrt{1 + 2\lambda(1 + a^2) + \lambda^2(1 - a^2)^2}}{2\lambda a} \tag{16}$$

the positive value of the root always being taken, so that

$$\lim_{\lambda \to \infty} \xi = \begin{cases} a & (|a| < 1) \\ a^{-1} & (|a| > 1) \end{cases} \tag{17}$$

as is otherwise clear from (7). The limiting values of $-\beta_1$ in the two cases are thus 0 and $a - a^{-1}$. That is, as regulation becomes more expensive, and so tends to the minimum possible consistent with stability, one finds two cases. If the system is intrinsically stable, then in the limit of infinitely costly regulation it will be unregulated. If it is intrinsically unstable, then a minimum amount of regulation will be required—this is, perhaps surprisingly, *not* the degree of regulation that only just secures stability, which would correspond to $\xi = 1-$ rather than a^{-1}. The point is that a greater degree of stabilisation will in fact be cheaper, because of the consequent decreased variability of all the series.

Ex. 1. Suppose that in the example above

$$\psi_\varepsilon = \frac{\sigma}{1 - \rho z}$$

with $|\rho| < 1$. Show that

$$\alpha(z) - \beta(z) = \frac{1 - \xi z}{1 - \Delta z}$$

where

$$\Delta = \frac{\rho(a - \xi)}{a(1 - \rho\xi)}$$

and hence that for optimum temperature regulation

$$x_t = -ay_{t-1} + (\xi - \Delta)\sum_{j=1}^{\infty} \Delta^{j-1} y_{t-j}$$

(ii) *A continuous time analogue of* (*i*); *demand and income stabilisation with lags*

Phillips (1958) has considered the stabilisation of income in a very simple model. We shall first consider one even simpler. Let

y = income or production per unit of time;
z = aggregate demand or sales;
v = policy demand, i.e. demand resulting from policy measures; introduced by the authorities in an attempt to stabilise y;
x = all demand except policy demand;
ε = random disturbance.

Suppose that

$$\frac{dy}{dt} = a(z - y) \tag{18}$$

$$z = x + v \tag{19}$$

$$x = cy + d + \varepsilon \tag{20}$$

Equation (18) states that production changes at a rate proportional to unsatisfied demand, while (20) relates non-policy demand in a static fashion to income.

There must be yet another relation, the "policy relation", expressing v in terms of past values of the system variables. Phillips in his paper considered a particular relation which he optimised with respect to a parameter. We shall use the methods of this chapter to calculate the unconstrained optimal relation under a variety of circumstances. We restrict ourselves still to linear rules, of course, so that the relation will be of the form

$$v = f + \int_k^\infty \beta_s y_{t-s}\, ds \tag{21}$$

for some $k \geqslant 0$.

A reasonable criterion might be

$$V = E[(y - \mu)^2 + \lambda v^2] \tag{22}$$

where μ is the desired income level.

As in (i), we readily find that the optimum levels of the mean values are given, if $E(\varepsilon) = 0$, by

$$\bar{y} = \bar{z} = \frac{\bar{x} - d}{c} = \frac{\bar{v} + d}{1 - c} = \frac{\mu + \lambda(1 - c)\, d}{1 + \lambda(1 - c)^2} \tag{23}$$

and these fix the value of the constant f.

In order to calculate the optimum filter β we need only consider fluctuations about mean values, these fluctuations being related by equations (18)–(22) if we set $d = f = \mu = 0$ (we shall economise on notation by denoting the fluctuation series by the same symbols as the "absolute" series: x, y, z and v).

We find, by elimination, that the regulation process is summed up in the equations

$$\frac{dy}{dt} + gy = a(v + \varepsilon) \tag{24}$$

$$v = \beta y = \int_k^\infty \beta_s y_{t-s}\, ds \tag{25}$$

where $g = a(1 - c)$. Since we shall expect that $0 < c < 1$, it will be true that $g > 0$, so the unregulated process is stable, at least.

The operator β is to be chosen so as to minimise

$$V' = E[y^2 + \lambda v^2] \tag{26}$$

This is just the problem treated in section (5) with $\alpha = (D + g)/a$, and, if $k = 0$, is precisely the continuous time analogue of the problem treated in (i). However, the assumption $k = 0$ is even less realistic in a continuous time model than in a discrete one, implying as it does that information on the regulated variable y is available immediately.

We have

$$\alpha(\omega) = \frac{1}{a}(g + i\omega) \tag{27}$$

$$|P|^2 = 1 + \lambda|\alpha|^2 = 1 + \frac{\lambda}{a^2}(g^2 + \omega^2) \tag{28}$$

so that P, the factor of $1 + \lambda|\alpha|^2$ which is analytic and zero free in the lower half-plane, is

$$P(\omega) = \sqrt{\frac{\lambda}{a^2}}(\xi + i\omega) \tag{29}$$

where

$$\xi = \sqrt{g^2 + \frac{a^2}{\lambda}} \tag{30}$$

Ex. 2. Show that if $k = 0$ and the process $\{\varepsilon_t\}$ consists of pure noise, so that $\psi_\varepsilon = \text{const.} = \sigma$ (say), then

$$\beta(\omega) = -\frac{a}{\lambda(\xi + g)} = -\frac{\xi - g}{a}$$

so that the control relation (25) is

$$v_t = -\left(\frac{\xi - g}{a}\right) y_t$$

Ex. 3. Show that if $k = 0$ and

$$\psi_\varepsilon(\omega) = \frac{\sigma}{\mu + i\omega}$$

then, under optimal running,

$$v_t = -\frac{\xi - g}{a(g + \mu)}\left[(\xi + g + \mu) + \frac{d}{dt}\right] y_t$$

This example shows the apparent difficulties one encounters if lags are not incorporated in the model: v_t depends upon y'_t, the very quantity v_t is determining. The simultaneity is best removed by regarding k as positive, although perhaps infinitesimally small.

Suppose we consider the case of a finite lag in information, k, under the supposition that $\psi_\varepsilon = \sigma$. Then, by formula (5.27)

$$\begin{aligned}
\phi &= \frac{1}{P}\left[\frac{P}{\alpha}\right]^{(k)*} + \frac{\lambda}{P}\left[\frac{\bar{\alpha}}{\bar{P}}\right]_k \\
&= \frac{1}{P}\left[\frac{P}{\alpha}\right]^{(k)*} \\
&= \frac{\alpha}{\xi + i\omega}\left[\frac{\xi + i\omega}{g + i\omega}\right]^{(k)*}
\end{aligned} \tag{31}$$

The asterisk indicates that the Fourier expansion must involve $e^{-ix\omega}$ for non-negative x only, i.e. it must be the expansion valid in some singularity-free region $Im(\omega) < \text{const.}$ In this particular case we can write

$$\frac{\xi + i\omega}{y + i\omega} = 1 + (\xi - g)\int_0^\infty e^{-(g+i\omega)x}\, dx \tag{32}$$

if $Im(\omega) < g$, so that

$$\left[\frac{\xi + i\omega}{g + i\omega}\right]^{(k)*} = \int_{0-}^{k} e^{-i\omega x}[\delta(x) + (\xi - g)e^{-gx}]\,dx$$

$$= 1 + \frac{\xi - g}{g + i\omega}(1 - e^{-k(g+i\omega)})$$

$$= \frac{\xi + i\omega}{g + i\omega} - \frac{\xi - g}{g + i\omega}\,e^{-k(g+i\omega)} \qquad (33)$$

Combining (31) and (33), we find then that

$$\beta(\omega) = \alpha - \phi^{-1}$$

$$= -\frac{\xi - g}{a}\,\frac{(g + i\omega)e^{-k(g+i\omega)}}{[(\xi + i\omega) - (\xi - g)e^{-k(g+i\omega)}]} \qquad (34)$$

The relation $v = \beta y$ can be written out explicitly from (34) if we remember that $i\omega$ corresponds to the operator $D = d/dt$, so that $e^{-ik\omega}$ corresponds to a lag of k.

Probably the best way of actually determining v in terms of y is to write the relation in the form

$$(\xi + D)v_t - (\xi - g)e^{-kg}v_{t-k} + \frac{\xi - g}{a}e^{-kg}(g + D)y_{t-k} = 0 \qquad (35)$$

If this is written

$$(\xi + D)v_t = R(t) \qquad (36)$$

then one could also consider using the recursion

$$v_t = \int_0^\infty e^{-\xi s}R(t - s)\,ds \qquad (37)$$

Ex. 4. Derive (34) by use of (5.28) rather than (5.27).

Ex. 5. Show that, for this last example

$$\text{var}\,(y) = \frac{\sigma^2 a^2}{2g}\left[1 - \left(\frac{\xi - g}{\xi}\right)e^{-2kg}\right]$$

$$\text{var}\,(v) = \frac{\sigma^2\,(\xi - g)^2\,e^{-2kg}}{2g}$$

$$(\text{var}\,(y) + \lambda\,\text{var}\,(v)) = \frac{\sigma^2 a^2}{2g}\left[1 - \frac{\xi - g}{\xi + g}\,e^{-2kg}\right]$$

As k increases from 0 to ∞, $2\,\text{var}\,(y)/(a\sigma)^2$ increases from $1/\xi$ to the unregulated value of $1/g$.

Ex. 6. It is interesting to calculate the amount of "policy variation" required to bring income variation down to a given specified value. Suppose we set

$$\text{var}\,(y) = \frac{\sigma^2 a^2}{2g}p$$

where p is a fixed value in the range (0, 1). Show then that

$$\text{var}\,(v) = \frac{(1 - p)^2}{p + e^{-2kg} - 1}$$

so that the required level of regulation is unachievable if

$$k \geqslant -\frac{1}{2g}\log\,(1 - p)$$

(iii) *Income stabilisation with inertia in the control variable*

Consider the model of (ii), and suppose that equations (18)–(20) continue to hold, but that v, the policy demand, cannot be determined directly. Instead, one can only determine a "potential policy demand" w, to which v is related by the equation

$$\frac{dv}{dt} = h(w - v) \tag{38}$$

The actual demand v will tend to follow w in a smoothed fashion. If we consider only fluctuations about mean values, then equations (24) and (25) will become modified to

$$\frac{dy}{dt} + gy = \frac{aw}{L} + a\varepsilon \tag{39}$$

$$w = \int_k^\infty \beta_s y_{t-s}\, ds \tag{40}$$

where

$$L = \frac{h + D}{h} \tag{41}$$

corresponding to

$$L(\omega) = \frac{h + i\omega}{h} \tag{42}$$

We shall suppose that the criterion quantity is

$$V = E(y^2 + \lambda w^2) \tag{43}$$

Since (39) may be rewritten

$$\frac{(h + D)(g + D)}{ah} y = w + \left(\frac{h + D}{h}\right)\varepsilon \tag{44}$$

the process is of the type already treated, with

$$\alpha(\omega) = \frac{(g + i\omega)(h + i\omega)}{ah} \tag{45}$$

and ψ_ε replaced by

$$\psi_\eta(\omega) = \frac{h + i\omega}{h}\psi_\varepsilon(\omega) \tag{46}$$

We shall suppose that $\psi_\varepsilon(\omega) = 1$, in which case (46) does not correspond to a proper process (cf. sections (3.6), (6.4)) but formal application of the method leads to correct results. We have

$$\begin{aligned} |P|^2 &= 1 + \lambda|\alpha|^2 \\ &= 1 + \lambda(g^2 + \omega^2)(h^2 + \omega^2)/(ah)^2 \\ &= \kappa^2(\xi_1{}^2 + \omega^2)(\xi_2{}^2 + \omega^2) \end{aligned} \tag{47}$$

say, where $Re(\xi_1, \xi_2) > 0$, so that

$$P = \kappa(\xi_1 + i\omega)(\xi_2 + i\omega) \tag{48}$$

We find that

$$\left[\frac{\psi\eta}{\alpha\bar{P}}\right]_k = \frac{a}{\kappa}\left[\frac{1}{(g + i\omega)(\xi_1 - i\omega)(\xi_2 - i\omega)}\right]_k$$

$$= \frac{a}{\kappa} \frac{e^{-k(g+iw)}}{(\xi_1 + g)(\xi_2 + g)(g + i\omega)} \tag{49}$$

Using (49), (5.28) and the readily verifiable fact that

$$\frac{1}{\kappa^2} = \frac{(ah)^2}{\lambda} = (\xi_1^2 - g^2)(\xi_2^2 - g^2) \tag{50}$$

we find that

$$\phi = \frac{ah}{(g + i\omega)(h + i\omega)}\left[1 - \frac{(\xi_1 - g)(\xi_2 - g)e^{-k(g+i\omega)}}{(\xi_1 + i\omega)(\xi_2 + i\omega)}\right] \tag{51}$$

$$\beta = -\frac{(g + i\omega)(h + i\omega)}{gh} \frac{(\xi_1 - g)(\xi_2 - g)e^{-k(g+i\omega)}}{(\xi_1 + i\omega)(\xi_2 + i\omega) - (\xi_1 - g)(\xi_2 - g)e^{-k(g+i\omega)}} \tag{52}$$

The relation $w = \beta y$ is best modified to a relation of the type (35) or (37). Note from (46), (51) and (52) that $y = \phi\eta$ and $w = \beta\phi\eta$ are proper (i.e. finite variance) processes.

Ex. 7. Show that ξ_1 and ξ_2 are conjugate complex quantities, so that the response of the system is damped oscillatory, if and only if

$$\lambda < \frac{4a^2h^2}{(g^2 - h^2)^2}$$

It is interesting to note that the response is always oscillatory if $h = g$, whatever the degree of regulation.

(iv) *The tracking example of Section 3, Ex. 1*

Consider the example of section 3, Ex. 1, with the modification that the future values of the quantity we are attempting to follow, u_t, are unknown, so that the relation of aerial position to input must be of the form

$$y_t = \theta u_t = \int_0^\infty \theta_s u_{t-s}\, ds \tag{53}$$

By (5.8) we have

$$\theta(\omega) = \frac{1}{\psi_u P}\left[\frac{\psi_u}{\bar{P}}\right]_+ \tag{54}$$

where

$$|P|^2 = 1 + \lambda|\alpha|^2 = 1 + \lambda M^2\omega^4 = 1 + \left(\frac{\omega}{f}\right)^4 \tag{55}$$

say. Thus

$$\begin{aligned} P &= (1 - \omega e^{-\pi i/4}/f)(1 - \omega e^{-3\pi i/4}/f) \\ &= \frac{1}{f^2}\left(i\omega + \frac{1+i}{\sqrt{2}}f\right)\left(i\omega + \frac{1-i}{\sqrt{2}}f\right) \\ &= \frac{1}{f^2}((i\omega)^2 + \sqrt{2}fi\omega + f^2) \end{aligned} \tag{56}$$

Suppose now that

$$\psi_u = \frac{\sigma}{\mu + i\omega} \qquad (\mu > 0) \tag{57}$$

Then we find

$$\left[\frac{\psi_u}{\bar{P}}\right]_+ = \frac{\psi_u(\omega)}{\bar{P}(i\mu)} = \frac{\psi_u(\omega)}{P(-i\mu)} \tag{58}$$

so that, by (54)

$$\theta(\omega) = \frac{1}{P(-i\mu)P(\omega)} = \frac{Af^2}{(i\omega)^2 + 2\zeta fi\omega + f^2} \tag{59}$$

where

$$A = \left[1 + \sqrt{2}\left(\frac{\mu}{f}\right) + \left(\frac{\mu}{f}\right)^2\right]^{-1} \tag{60}$$

$$\zeta = \frac{1}{\sqrt{2}} \tag{61}$$

Equation (59) gives the response function of the system as a whole; the corresponding weighting function is

$$\theta_s = \frac{1}{2\pi}\int e^{i\omega s}\theta(\omega)\, d\omega = \sqrt{2}Afe^{-fs/\sqrt{2}} \sin\left(\frac{fs}{\sqrt{2}}\right) \tag{62}$$

The response to a unit step function is thus given by

$$\begin{aligned} R(t) = \int_0^t \theta_s\, ds &= A\left[1 - e^{-ft/\sqrt{2}}\left(\cos\left(\frac{ft}{\sqrt{2}}\right) + \sin\left(\frac{ft}{\sqrt{2}}\right)\right)\right] \\ &\longrightarrow A \end{aligned} \tag{63}$$

From (60) and (63) we see that there is a static lag, which tends to zero with μ, as we should expect from the results on accumulated process inputs in section (6).

In cquation (59) we have expressed the frequency response of the system in a standard form; the parameter ζ determines the amount of overshoot in the time response to the unit step function. From Fig. 1.18 of Truxal (1955) we see that a ζ value of 0·707 corresponds to an overshoot of about 4%. This is very moderate, and would seem to contradict the often-made statement that systems based on least-square criteria have a tendency to be underdamped, by conventional standards.

Ex. 8.

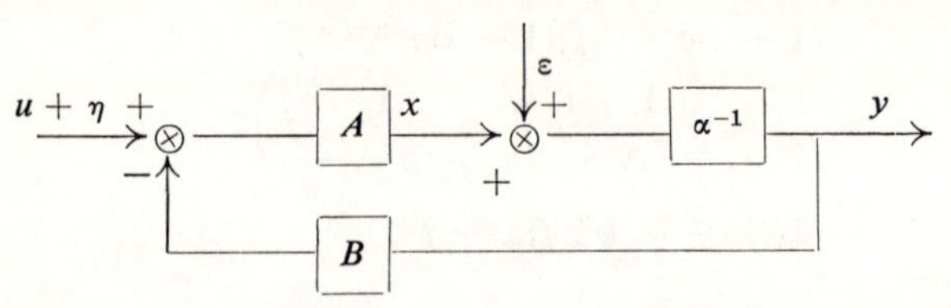

Block diagram of the regulating system considered in Exercises 8 to 13.

A more general tracking example is illustrated: the input u is corrupted by an error η; $A(\omega)$ is the response of the servo-amplifier and filter, α^{-1} that of servo-motor, gear train, and aerial, B that of filter in the feed-back loop. The error, ε, is that due to wind torques on aerial, etc.

The α response is given. If we try to minimise $E[(y - u)^2 + \lambda x^2]$ by variation of A and B we recover essentially the disturbed following problem of section 5 (ii). However, a more conventional approach is to set $B = 1$, and choose A so as to minimise $E(y - u)^2$. Judged by our previous considerations this seems somewhat arbitrary, in that the rate of working of the servo-motor is not limited explicitly, but by virtue of a special feature of the design—the feed-back loop is made simple.

We shall accept the limitation $B = 1$, but, for the moment, retain the criterion $E[(y - u)^2 + \lambda x^2]$.

Show that if the processes $\{u\}$, $\{\eta\}$ and $\{\varepsilon\}$ are mutually uncorrelated with s.d.f.'s $f_{uu}(\omega)$, etc., and

$$\theta = \left(\frac{y}{u + \eta}\right) = \frac{A}{A + \alpha}$$

$$|P|^2 = \left[f_{uu} + f_{\eta\eta} + \frac{f_{\varepsilon\varepsilon}}{|\alpha|^2}\right][1 + \lambda|\alpha|^2]$$

then the optimum physically realisable operator A will be that for which

$$\theta = \frac{1}{P}\left[\frac{f_{uu} + f_{\varepsilon\varepsilon}/|\alpha|^2}{\bar{P}}\right]_+$$

(The presence of the term $|\alpha|^{-2}$ is objectionable if α is singular, as is usually the case. The simplest course is to regard α as the limit of a non-singular physically realisable operator (see Ex. 10 below): the limit values of A and θ are always sensible if the problem admits a proper solution at all.)

Ex. 9. Show that if $\varepsilon_t \equiv 0$, $\lambda = 0$ in Ex. 8, then the whole system is equivalent to the Wiener filter for the separation of signal from noise (cf., e.g., (6.1.13)).

Ex. 10. Suppose that in Ex. 8

$$\alpha(\omega) = (k + i\omega) \qquad (k > 0)$$

$$f_{uu} = \frac{\sigma^2}{\omega^2 + \mu^2} \qquad (\mu > 0)$$

and $f_{\eta\eta}$, $f_{\varepsilon\varepsilon}$ are constants. Show that if $f_{\varepsilon\varepsilon} = 0$ then

$$\theta(\omega) = \frac{(\rho - \mu)(\nu - \mu)}{(\rho + i\omega)(\nu + i\omega)}$$

$$A(\omega) = \frac{(k + i\omega)(\rho - \mu)(\nu - \mu)}{(\rho + i\omega)(\nu + i\omega) - (\rho - \mu)(\nu - \mu)}$$

where

$$\rho = \sqrt{\mu^2 + \frac{\sigma^2}{f_{\eta\eta}}}$$

$$\nu = \sqrt{\frac{1 + \lambda k^2}{\lambda}}$$

Show that, if, on the other hand, $\lambda = 0$ ($f_{\varepsilon\varepsilon}$ not necessarily being zero), then

$$\theta(\omega) = 1 - \frac{(k + i\omega)(\mu + i\omega)}{(\alpha + i\omega)(\beta + i\omega)}$$

$$A(\omega) = \frac{(\alpha + i\omega)(\beta + i\omega) - (\mu + i\omega)(k + i\omega)}{\mu + i\omega}$$

where α, β have positive real part and $[f_{\eta\eta}(k^2 + \omega^2)(\mu^2 + \omega^2) + \sigma^2(k^2 + \omega^2) + f_{\epsilon\epsilon}(\mu^2 + \omega^2)]$ is zero for $\omega^2 = -\alpha^2, -\beta^2$. Note that in both cases the limit system for $k \downarrow 0$ is a perfectly proper one.

Ex. 11. By appealing to the theorems of section (8.5), show that if $f_{\epsilon\epsilon} \neq 0$ in Ex. 8, and $\alpha(\omega)$ tends to $c(i\omega)^p$, then

$$\theta(\omega) = 1 + O(\omega^p)$$

Ex. 12. Suppose that the input is a simply accumulated process, so that we consider

$$f_{uu}(\omega) = \frac{f_{vv}(\omega)}{\omega^2 + \mu^2}$$

in the limiting case $\mu \downarrow 0$. Show that, as in section (6), the system will behave properly if, and only if, $\alpha(\omega)$ is $O(\omega)$ at the origin, and that then

$$\theta(\omega) = 1 + O(\omega)$$

If $\alpha(\omega)$ is of order ω exactly (i.e. not of higher order) show that

$$\theta = \frac{1}{R}\left\{\left[\frac{f_{vv} + \omega^2 f_{\epsilon\epsilon}/|\alpha|^2}{\bar{R}}\right]_+ + \left(\left[\frac{f_{vv} + \omega^2 f_{\epsilon\epsilon}/|\alpha|^2}{\bar{R}}\right]_-\right)_{\omega=0}\right\}$$

where

$$|R|^2 = [f_{vv} + \omega^2 f_{\eta\eta} + \omega^2 f_{\epsilon\epsilon}/|\alpha|^2][1 + \lambda|\alpha|^2]$$

Ex. 13. Generalise Ex. 12 to accumulated inputs of arbitrary order.

Ex. 14. Referring again to the production example of section (2), suppose that the quantity we wish to minimise is

$$V = E[A_1(P_t - K_1)^2 + A_2(P_t - P_{t-1})^2 + A_3(I_t - K_2)^2]$$

where one has the exact relation between production, demand and inventory,

$$P_t - S_t = I_t - I_{t-1}$$

and the decision relation is to be of the form

$$I_t = \bar{I} + \beta S = \bar{I} + \sum_k^{\infty} \beta_j S_{t-j}$$

where $k \geqslant 0$.

Show that the optimum values are given by

$$\bar{I} = K_2 - \beta(1)E(S)$$

and

$$\beta = -\frac{1}{\psi_s R}\left[\frac{\psi_s Q}{\bar{R}}\right]_k$$

where $|\psi_s|^2$ is s.d.f. of $\{S_t\}$,

$$|R|^2 = A_1|1 - z|^2 + A_2|1 - z|^4 + A_3$$

$$Q = A_1(1 - z^{-1}) + A_2(1 - z)(1 - z^{-1})^2$$

Ex. 15. If, in the previous example, $A_2 = 0$ and

$$\psi_s = \frac{\sigma}{1 - \rho z} \qquad (|\rho| < 1)$$

show that

$$\left(\frac{I}{S}\right) = \beta = -\frac{\xi(1-\rho)}{1-\rho\xi}\frac{\rho^k}{1-\xi z}$$

and that if $k = 0$, then

$$\left(\frac{P}{S}\right) = \frac{1-\xi}{1-\rho\xi}\frac{1-\rho\xi z}{1-\xi z}$$

where ξ^{-1} is the root of R. Note that this latter transfer function has value unity for $z = 1$: interpret.

Ex. 16. Suppose that in Ex. 13 $A_1 = 0$, corresponding to the assumption that production costs are linear in P. If $k = 0$, ψ_s has the form given in Ex. 14, and ξ^{-1}, η^{-1} are the roots of R, show that

$$\beta = \frac{\xi\eta(1-\rho)^3}{\rho(1-\rho\xi)(1-\rho\eta)}\frac{1}{(1-\xi z)(1-\eta z)} - \frac{\xi\eta}{\rho}\frac{1-\rho z}{(1-\xi z)(1-\eta z)}$$

Show also that ξ, η are complex conjugates, and that if

$$\xi, \eta = e^{-\lambda \pm i\mu}$$

then

$$\cosh(\lambda)\cos(\mu) = 1$$

These complex roots correspond to a damped periodic response, caused by the $(P_t - P_{t-1})^2$ term in the cost function, which introduces a degree of inflexibility into production changes, with consequent overshoot.

Ex. 17. Consider again the production work-force example with expected cost given by equation (2.12). This problem was solved in section (3) for the case of known future demand, but suppose that the decision relations are now restricted to the form

$$W_t = \text{const.} + \beta S_t = \text{const.} + \sum_0^\infty \beta_j S_{t-j}$$

$$I_t = \text{const.} + \gamma S_t = \text{const.} + \sum_0^\infty \gamma_j S_{t-j}$$

If the equation system (3.9) is written in matrix form

$$\mathbf{B}(z)\begin{pmatrix}\beta\\ \gamma\end{pmatrix} = \mathbf{C}(z) \qquad (64)$$

show that the solution to the restricted problem is

$$\begin{pmatrix}\beta\\ \gamma\end{pmatrix} = (\psi_s\mathbf{R})^{-1}[\psi_s(\mathbf{R}^\dagger)^{-1}\mathbf{C}]_+ \qquad (65)$$

where $|\psi_s|^2$ is the s.d.f. of $\{S_t\}$ and

$$\mathbf{B} = \mathbf{R}^\dagger\mathbf{R} \qquad (66)$$

the matrices $\mathbf{R}$ and $\mathbf{R}^{-1}$ having elements regular in $|z| \leqslant 1$.

The factorisation of **B** can be carried out by the methods of section (9.3). An alternative method which, superficially at least, avoids factorisation, is given in the next section.

We have had earlier examples in which it was necessary to optimise more than one filter, but the function to be factorised was always a scalar. The present example is typical of a wider class of problems.

Ex. 18. Consider the stabilisation of the simple "lagged accelerator" model, (Tustin (1953), p. 16) in discrete time:

$$I = \alpha Y$$

$$Y = (\beta + \gamma)I + \varepsilon$$

where $Y =$ income, $I =$ rate of decisions to invest, α has the nature of a lagged difference, γ the nature of a "distribution", and β the stabilising operator, to be determined, reflecting government action. All operators are one-sided. The "policy" term is βI, so if all quantities are measured from equilibrium values a reasonable criterion function would be

$$V = E[Y^2 + \lambda(\beta I)^2]$$

Find the general solution, and that for the particular case

$$\alpha = \alpha_0 z^k(1 - \xi z) \qquad \text{(where } k > 0\text{)}$$

$$\gamma = \frac{\beta_0}{(1 - \rho z)^2}$$

$$f_{\varepsilon\varepsilon} = \text{const.}$$

10.8 Certainty Equivalence

We saw in section (4) that, when there is no uncertainty in the futures of the input series ("errors" being classed as "inputs" in this connection), then in a wide class of cases one will obtain equivalent control rules if one minimises a statistically averaged criterion function or a certain quadratic form, the time averaged criterion function. This conclusion must certainly be modified if the inputs have an uncertain future. However, the modification turns out to be a rather simple and attractive one: roughly, one substitutes least-square estimates of all variables in the criterion function which have not yet been observed, and one then chooses the current value of the control variables to minimise this quadratic form, just as if the estimated values were the true values. This principle is known as "certainty equivalence"; it seems to be due originally to Simon (1956) and has since been generalised by Theil (1957).

We shall now prove an extended form of the principle. We shall assume all series of random variables to be purely non-deterministic, for by the argument of Theorem (4.4.2) or equation (6.3), the deterministic components can be considered separately, and present no real problem.

Theorem

Suppose the control variables x_t are to be chosen so as to minimise

$$V = EQ(x_1, x_2, \ldots x_n; u_1, u_2, \ldots u_n; v_1, v_2, \ldots v_n)$$

where Q is a positive definite quadratic form. The series $\{u\}$, $\{v\}$ are sequences of random variables whose unconditional means are zero, and x_t is restricted to being a linear function of $u_1, u_2, \ldots u_t$. Then, $x_1, x_2, \ldots x_{t-1}$ having been determined, x_t should be determined by minimising the quadratic form

$$Q(x_1 \ldots x_n; u_1 \ldots u_t, u_{t+1}^{(t)} \ldots u_n^{(t)}; v_1^{(t)} \ldots v_n^{(t)})$$

with respect to $x_t, x_{t+1} \ldots x_n$. Here, $u_j^{(t)}$, $v_j^{(t)}$ are the l.l.s.e. of u_j and v_j in terms of $u_1, u_2, \ldots u_t$, and $x_1, x_2, \ldots x_n$.

Here the u's should be understood as describing all information currently available (i.e. in the aerial tracking problem: observed position of target, observed position of aerial (which may not be the same as either the actual or the intended position), etc.). The v's describe unobservables, such as error in the observation of the target.

It will be noted that only a finite time interval is considered, $t = 1, 2, \ldots n$. One would expect the results to extend to the infinite interval situation, at least under mild conditions, but we shall not attempt a proof of this fact.

In determining x_t one also obtains determinations of later x values, which we shall denote by $x_j^{(t)}$ $(j > t)$. These should be regarded as preliminary determinations of future control values, which will be revised as new information $(u_{t+1}, u_{t+2}, \ldots)$ becomes available.

To prove the theorem: successively orthogonalise the inputs u_j to obtain a series of innovations $\varepsilon_1, \varepsilon_2, \ldots$ (cf. equations (4.2.5), (4.2.6)). For simplicity we shall assume these to be scalar variates, but the proof generalises immediately to the vector case. Since x_t is a linear function of $u_1, u_2, \ldots u_t$ there will be a representation

$$x_t = \sum_{s=1}^{t} b_{st}\varepsilon_s \tag{1}$$

Let us set

$$x_j^{(t)} = \sum_{s=1}^{\min(j,\, t)} b_{sj}\varepsilon_s \tag{2}$$

so that

$$x_j = x_j^{(t)} + \tilde{x}_j^{(t)} \tag{3}$$

say, where $\tilde{x}_j^{(t)}$ is uncorrelated with $u_1, u_2, \ldots u_t$. We have not proved that $x_j^{(t)}$ as defined by (2) is identical with the quantity defined in the last paragraph (if $j > t$), but it will become clear that this is true.

As in the proof of Theorem (4.4.1), it is then evident that

$$\underset{\substack{u_1 \ldots u_n \\ v_1 \ldots v_n}}{E} Q(x, u, v) = \text{const.} + \underset{u_1 \ldots u_t}{E} Q(x^{(t)}, u^{(t)}, v^{(t)}) \tag{4}$$

in an obvious notation. Since $x_j^{(t)}$ $(j \geqslant t)$ is unrestricted as a function of any of the other arguments of $Q(x^{(t)}, u^{(t)}, v^{(t)})$, we can choose the values which maximise this quadratic form freely, as in Theorem (4.4.1). The theorem is thus proved.

It might be interesting to verify directly on a particular example of the type treated in sections (5) and (7) that the two methods are equivalent.

Consider the choosing of $\{y_t\}$ so as to minimise

$$V = E[(y_t - u_t)^2 + \lambda(\alpha y_t)^2] \tag{5}$$

where α is an operator into the past, and $\{u_t\}$ a stationary input process with s.d.f. $|\psi_u|^2$ and an unknown future. The variable y can be regarded as equivalent to the control variable, x, if there is no error in control. If y is to be calculated in the form

$$y_t = \sum_0^\infty \theta_j u_{t-j}$$

then $\theta(z)$ is determined by equations (5.8), (5.9).

Working from the certainty equivalence method, we see that y_t should be calculated from the equation system

$$|P|^2 y_{t+s}^{(t)} = u_{t+s}^{(t)} \qquad (s \geqslant 0) \tag{6}$$

where $|P|^2$ operates on the subscript s in (6) instead of t, $u_{t+s}^{(t)}$ is the linear least-square predictor of u_{t+s} calculated from $u_t, u_{t-1}, u_{t-2}, \ldots$, and

$$y_{t+s}^{(t)} = y_{t+s} \qquad (s \leqslant 0) \tag{7}$$

Now seeing that (6) holds for $s \geqslant 0$, we can apply the operator $\bar{P}^{-1}$, which operates only into the future, and obtain the relation

$$P y_{t+s} = \bar{P}^{-1} u_{t+s}^{(t)} \qquad (s = 0) \tag{8}$$

We thus have a relation determining y_t, from which the provisional estimates $y_{t+s}^{(t)}$ $(s > 0)$ have been eliminated.

Now

$$u_t = \psi_u \varepsilon_t \tag{9}$$

where $\{\varepsilon_t\}$ is a sequence of uncorrelated standardised variates, and further

$$u_{t+s}^{(t)} = \psi_u \varepsilon_{t+s}^{(t)} \qquad (s > 0) \tag{10}$$

where ψ_u now operates on the s subscript, and

$$\varepsilon_{t+s}^{(t)} = \begin{cases} 0 & (s > 0) \\ \varepsilon_{t+s} & (s \leqslant 0) \end{cases} \tag{11}$$

We thus see that (8) can be written

$$P y_t = [\bar{P}^{-1}\psi_u]_+ \varepsilon_t = [\bar{P}^{-1}\psi_u]_+ \psi_u^{-1} u_t \tag{12}$$

in agreement with (5.8).

As a practical procedure, the method will be advantageous only if it is easier to solve the equation system (6) numerically (or its equivalent in more general cases) than to make use of the literal solution determined by (5.8). For every time instant t one would have to solve an infinite equation system in order to calculate the single value y_t; not a light matter. However, sometimes one would lose little efficiency by working with a finite time-horizon in the future, in which case the system (6) would be finite. There are also cases in practice where one can forecast the input series, u_t, not by least-square methods, but from miscellaneous data and from personal experience. By inserting this semi-subjective forecast in (8) one can achieve some reasonable degree of regulation, although it is impossible to say how efficient such a procedure would be.

For some multivariate examples a direct use of (6) rather than of (5.8) has the advantage that one does not need to factorise an operator matrix, as we shall see in an example below, although one has a problem falling not far short of factorisation.

If regulation is to be achieved by mechanisms rather than by computation, then one has no other alternative than to obtain the solution (5.8) in literal form, when designing the mechanism.

Ex. 1. (Holt *et al.*, 1960.) Suppose that

$$Y_t = I_{t-1} + S_{t-1} + C_{t-1}$$
$$C_t = cY_t$$

where

$Y =$ national income;
$C =$ consumer spending;
$S =$ government spending;
$I =$ private investment (s.d.f. $|\psi_I|^2$);
$c =$ marginal propensity to consume;

and suppose that government spending is to be determined by the relation of the type

$$S_t = \mu + \beta I_t = \mu + \sum_1^{\infty} \beta_j I_{t-j}$$

which minimises

$$E[(Y - Y^*)^2 + \lambda(S - S^*)^2]$$

where Y^*, S^* are prescribed "desirable" levels. Show that our previous methods give

$$\beta = -\frac{1 - cz}{\psi_I P}\left[\frac{\psi_I}{(1 - cz)\bar{P}}\right]$$

$$|P|^2 = 1 + \lambda|1 - cz|^2 = K(1 - \xi z)(1 - \xi z^{-1})$$

say, where $|\xi| < 1$, while the certainty equivalence method gives

$$S_t = \text{const.} - \frac{1}{K(1 - c\xi)}\left[\sum_0^{\infty} \xi^j I_{t+j}^{(t-1)} + cY_t\right]$$

where $I_{t+j}^{(t-1)}$ is the l.l.s.e. of I_{t+j} based on $I_{t-1}, I_{t-2} \ldots$

Example

The production work-force example associated with relation (2.12) had a solution given by (3.9), or (7.64), equivalent to an equation system

$$\mathbf{B}(U)\begin{bmatrix} W_t \\ I_t \end{bmatrix} = \mathbf{C}(U)S_t \tag{13}$$

The solution for unknown future S_t, given by (7.65), requires a canonical factorisation of the operator matrix $\mathbf{B}$ of the type (7.66).

The certainty equivalence method seems to avoid this factorisation, because it states that the equation system (13) will continue to hold if $S_{t+s}^{(t)}$ is substituted for S_{t+s} $(s > 0)$.

However, one has still to deal with a semi-infinite equation system, and, as in the passage from (6) to (8), one cannot avoid a calculation which amounts to a canonical factorisation.

10.9 General Comments on Least-square Regulation

It has been said that the mean square criterion attaches excessive weight to large deviations, relative to medium ones, with the consequence that systems designed on the least-square principle have over-hasty correction for deviation. This is said to produce a tendency to oscillation, sometimes of the rapid, low-amplitude "flutter" type.

It must certainly often be true that the mean-square criterion is not ideal, and that it may exaggerate the relative importance of large deviations. However, this in itself should not produce instabilities of the type mentioned. We noted after equation (7.63) that the least-square system for making a massive, mechanically undamped aerial follow a Markovian input signal showed a response to a unit step function with a maximum overshoot of 4%, whatever the rate of working of the turning force, or the predictability of the input. An overshoot as low as 4% indicates a very conservative rate of correction, and heavy damping of oscillations.

The probable reason for the observed instabilities is that account has not been taken, in the criterion function and in the design, of all "noise-sources" in the system. A precise method of optimisation based on a precise model of the system will take full advantage of any special features of this model, such as stationarity of inputs, or absence of internally generated error at various points in the system. If some of these "special features" have found their way in to the model by negligence rather than by design, then the performance of the system will suffer accordingly. Like a computer, a mechanical method of optimisation will work on from one's instructions very literally, and one must take care to allow for all contingencies in these "instructions" (specification of model and criterion).

Then, again, the criterion function should be sensitive to the specific types of deviation one wishes to avoid. If one objects to low-amplitude "jittery" motions, then obviously one is sensitive, not only to the amplitude of the deviations, but also to the amplitude of their derivatives, and the criterion function should be similarly sensitive.

To sum up: mechanical methods of design can be no substitute for insight, care and sophistication, but they should enable one to make the most of these qualities.

REFERENCES

Bartlett, M. S. (1955) *An Introduction to Stochastic Processes.* Cambridge University Press.

Bharucha-Reid, A. T. (1960) *Elements of the Theory of Markov Processes and their Applications.* McGraw-Hill, New York.

Bellman, R. (1957) *Dynamic Programming.* Princeton University Press.

Bellman, R. (1961) *Adaptive Control Processes: A Guided Tour.* Princeton University Press.

Cox, D. R. (1961) "Prediction by exponentially weighted moving averages and related methods", *J. Roy. Statist. Soc., B.*, **34**, 414–422.

Cramér, H. (1942) "On harmonic analysis in certain functional spaces", *Ark. mat. astr. fys.*, **28**, No. 12.

Doob, J. L. (1953) *Stochastic Processes.* Wiley, New York.

Durbin, J. (1960) "The fitting of time-series models", *Rev. Int. Inst. Stat.*, **28**, 233–244.

Furstenberg, H. (1960) *Stationary Processes and Prediction Theory.* Princeton University Press.

Grenander, U. (1954) "On the estimation of regression coefficients in the case of an autocorrelated residual", *Ann. Math. Stat.*, **25**, 252–272.

Grenander, U., and Rosenblatt, M. (1957) *Statistical Analysis of Stationary Time Series.* Wiley, New York.

Hartree, D. R. (1952) *Numerical Analysis.* Oxford.

Helson, H., and Lowdenslager, D. (1958) "Prediction theory and Fourier series in several variables", *Acta Math.*, **99**, 165–202.

Holt, C., Modigliani, F., Muth, J. F., and Simon, H. A. (1960) *Planning, Production, Inventories and Work-force.* Prentice-Hall.

James, H. M., Nichols, N. B., and Phillips, R. S. (1947) *Theory of Servomechanisms.* McGraw-Hill.

Kendall, D. G. (1949) "Stochastic processes and population growth", *J.R. Statist. Soc. B*, **11**, 230.

Kolmogorov, A. (1939) "Sur l'interpolation et l'extrapolation des suites stationnaires", *C.R. Acad. Sci. Paris*, **208**, 2043–2045.

Kolmogorov, A. (1941a) "Stationary sequences in Hilbert space", *Bull. Math. Univ. Moscou* 2, *no.* 6.

Kolmogorov, A. (1941b) "Interpolation und Extrapolation von stationären zufälligen Folgen", *Bull. Acad. Sci. (Nauk) U.R.S.S., Ser. Math.*, **5**, 3–14.

Laning, J. H., and Battin, R. H. (1956) *Random Processes in Automatic Control.* McGraw-Hill, New York.

Leslie, P. H. (1945) "On the use of matrices in certain population mathematics", *Biometrika*, **33**, 183–212.

Leslie, P. H. (1948) "Some further notes on the use of matrices in population mathematics", *Biometrika*, **35**, 213–245.

Newton, G. C. (1952) "Compensation of feedback control systems subject to saturation", *J. Franklin Inst.*, **254**, 281–286, 391–413.

Newton, G. C., Gould, L. A., and Kaiser, J. F. (1957) *Analytical Design of Linear Feedback Controls*, Wiley, New York.

Noble, B. (1958) *The Wiener–Hopf Technique*. Pergamon Press, London.

Peterson, E. L. (1961) *Statistical Analysis and Optimisation of Systems*. Wiley, New York.

Phillips, A. W. (1958) "La cybernétique et le contrôle des systèmes économiques", *Cahiers de l'Institut de Science Économique Appliquée. Série N*, **No. 2**, 41–48.

Rosenblatt, M. (1957) "A multidimensional prediction problem", *Ark. f. Mat.*, **3**, 407–424.

Simon, H. A. (1956) "Dynamic programming under uncertainty with a quadratic criterion function", *Econometrica*, **24**, 74–81.

Solodovnikov, V. V. (1960) *Introduction to the Statistical Dynamics of Automatic Control Systems*. Dover, New York (Russian original, 1952).

Theil, H. (1957) "A note on certainty equivalence in dynamic planning", *Econometrica*, **25**, 346–349.

Thompson, P. D. (1956) "Optimum smoothing of two-dimensional fields", *Tellus*, **8**, 384–393.

Truxal, J. G. (1955) *Automatic Feedback Control System Synthesis*, McGraw-Hill, New York.

Tsien, H. S. (1954) *Engineering Cybernetics*. McGraw-Hill, New York.

Tustin, A. (1953) *The Mechanism of Economic Systems*. Heinemann, London.

Whittle, P. (1952) "The simultaneous estimation of a time series' harmonic and covariance structure", *Trab. Estad.*, **3**, 43–57.

Whittle, P. (1954a) "On stationary processes in the plane", *Biometrika*, **41**, 434–449.

Whittle, P. (1954b) "The statistical analysis of a seiche record", *J. Marine Res.*, **13**, 76–100.

Wiener, N. (1949) *Extrapolation, Interpolation and Smoothing of Stationary Time Series*. Wiley, New York.

Wiener, N., and Masani, P. (1957, 1958) "The prediction theory of multivariate stochastic processes, I–II", *Acta Math.*, **98**, 1–2, 111–150; **99**, 1–2, 93–137.

Wilks, S. S. (1962) *Mathematical Statistics*. Wiley, New York.

Wold, H. (1938, 2nd ed., 1954) *The Analysis of Stationary Time Series*. Almquist and Wicksell, Uppsala.

Zadeh, L. A., and Ragazzani, R. (1950) "Extension of Wiener's theory of prediction", *J. Appl. Phys.*, **21**, 7, 645–655.

Yaglom, A. M. (1955) "The correlation theory of processes whose *n*th differences constitute a stationary process", *Matem. Sb.*, **37** (79) 1, 141–196.

Yaglom, A. M. (1960) "Effective solutions of linear approximation problems for multivariate stationary processes with rational spectrum", *Teoriya Veroyatnostei*, **5**, 265–292.

NAME AND SUBJECT INDEX